AF564824

A Colour Handbook on Rainfed Rabi Crops

A Colour Handbook on Rainfed Rabi Crops

Protection, Constraints and Mitigation Strategies

Authors

Reena
Sonika Jamwal
Anil Kumar
Jai Kumar
P.K. Rai
Brinder Singh
R. Puniya

NEW INDIA PUBLISHING AGENCY
New Delhi – 110 034

NEW INDIA PUBLISHING AGENCY
101, Vikas Surya Plaza, CU Block, LSC Market
Pitam Pura, New Delhi 110 034, India
Phone: + 91 (11) 27 34 17 17 Fax: + 91 (11) 27 34 16 16
Email: info@nipabooks.com
Web: www.nipabooks.com

Feedback at feedbacks@nipabooks.com

ISBN: 978-93-86546-69-2

Composed, Designed and Printed in India

Foreword

Intensive cultivation, monocropping and excessive indiscriminate use of chemical pesticides and fertilizers have not only damaged the entire eco-system, but is also responsible for the rise in several pest incidences in the recent past. Well known examples are of whitefly attack in cotton, mealy bug havoc in several crops, leaf folder attack in rice, etc. In the developing countries, the problem of competition from insect pests is further complicated with a rapid annual increase in the human population (2.5-3.0 percentage) in comparison to a 1.0 percentage increase in food production. To overcome this problem, we need to educate mainly the farmers and agriculture department field functionaries, who are in direct contact with the farmers. In field, exact and correct identification of the field problem, based on the symptoms, is of prime importance. The symptoms are usually very confusing in the field and not only farmers but also the field functionaries, sometimes gets confused with the correct identification, thus resulting in wrong pesticide application.

This book 'A Colour Handbook: Crop Protection of Rain Fed Rabi Crops' provides pictorial depictions of damaging symptoms and the pest itself along with their mitigation strategies in a very simple and concise manner. It is a very timely effort to help the farming community and reduce the pesticidal load in the environment. It shall be of immense use to UG and PG students along with other stake holders interested in exact identification and management of pests.

I congratulate the authors for bringing out this publication at an appropriate time, taking great efforts.

JPSharma.

Dr. Jag Paul Sharma
Director Research
SKUAST-Jammu

Date: 10/10/2017
Place: Jammu

Preface

Without preventive protection with pesticides, natural enemies, host plant resistance and other nonchemical controls, 70% of crops could have been lost to pests. Both abiotic and biotic factors are responsible for crop losses. Among the biotic factors, weeds cause the highest potential loss (30%), with animal pests and pathogens being less important (losses of 23 and 17%). Crop protection techniques have been developed for the prevention and control of crop losses due to pests in the field. However, the efficacy of control of pathogens and animal pests only reaches 32 and 39%, respectively, compared to almost 74% for weed control. Besides, less than 1% of the pesticides applied actually reaches the target pest. Rest more than 99% polluted our environment.

Insects are the most diverse species of animals living on earth. Providing food has always been a challenge facing mankind. A major bottleneck in this challenge is the competition from insect pests. Particularly in the tropics and sub-tropics, where the climate provides a highly favourable environment for a wide range of insects, diseases, etc., massive efforts are required to suppress population densities of different pests in order to achieve an adequate supply of food.

Rabi crops of rainfed areas are inflicted by several important insect pests, diseases and weeds. Correct identification of insect pests, diseases and weeds is therefore necessary not only for strict quarantine to check the spread of new pest species, but to achieve the desired productivity levels. The purpose of this publication entitled 'A Colour Handbook on Rainfed Rabi Crops Protection, Constraints and Mitigation Strategies' is to assist the students, field researchers, agriculture department field functionaries, scholars and farmers in correct identification of these pests and their management thereof. In this book, efforts have been made to describe the damaging symptoms, identification characteristics, biology and management techniques of more than 100 pest species with their coloured illustrations.

We would like to extend my gratitude to all contributors for their up to date scientific information organized in a befitting manner. The cooperation extended by Dr. J.P. Sharma, Director Research of the University is duly acknowledged. Thanks are also due to other supporting staff members of ACRA-Dhiansar, Bari Brahmana, SKUAST-J for their helping hand in successful completion of this book.

December, 2017 **Authors**

Contents

Section B: Major Diseases

Section D: Plant Nutrient Functions, their Deficiency and Toxicity Symptoms in Rainfed Rabi Crops

Section - A
Major Insect Pests of Rainfed Rabi Crops

1

Cereals

Wheat/Barley

1. Wheat aphid, *Macrosiphum miscanthi* (Hemiptera; Aphididae)

Distribution

It attacks wheat, barley, oats, etc. and is widely distributed in India.

Damaging symptoms

Both nymphs and adults suck sap from the ears and tender leaves. Leaves become pale and the yield goes down.

Life cycle

The wheat aphid breeds at faster rate during cold weather and reaches its peak during February – March, when the ears start ripening. The females give birth to young nymphs (viviparous), which undergo a number of moults and become adults in 7-8 days period. During winters, they reproduce parthenogenetically, but during summers, the winged males and females are produced, which migrate to other host plants.

Management strategy

- Installation of yellow sticky traps @ 50 per ha.
- Spray systemic insecticides Dimethoate 30EC or Imidacloprid 17.8 SL @ 2 ml / L.
- Destroy the affected parts along with aphid population in the initial stage.
- Conservation of the natural enemies: Ladybird beetles viz., *Coccinella septempunctata, Menochilus sexmaculata, Hippodamia variegata and Cheilomones vicina* are most efficient pradators of the mustard aphid. Adult beetles may feed on an average of 10 to 15 adults /day.
- Sprayed to manage the aphid population.

Wheat aphid nymphs and adults

Lady bird beetle, *Coccinella* sp. grub, pupa and adult

2. Armyworm, *Mythimna separata* (Lepidoptera: Noctuidae)

Distribution

The pest is present all over the world. In India it is a sporadic pest of wheat, sugarcane, maize, jowar, bajra, etc.

Damaging symptoms

Early larval instars feed on tender leaves, while the later instars are voracious feeders. They completely skeletonize the plants, giving a grazed look to the fields.

Life cycle

The pale brown to brick red coloured adult moths, lay round, light green eggs (100 in no.) singly in rows or in clusters on dry or fresh plants or on the soil. The eggs later turn pale yellow and then finally black and hatch in 4 – 13 days. There are six larval instars, which pupate in 14-21 days in soil at a depth of 0.5 – 5.0 cm or in dried leaves or stubbles. The pupation period is of 7 – 30 days. With the rise in temperature during March – April adult population builds up.

Management strategy

- Hand collection and destruction of larvae.
- Deep summer ploughing to expose the pupae to natural enemies like predatory birds.
- Spraying with Carbaryl 50 WP or Dichlorvos 100 EC or Trichlorfon 5% GR/5% Dust/ 50% EC or Methylparathion 50% EC @ 2 ml/L of water.

Armyworm larva and adult

3. Ghujhia weevil, *Tanymecus indicus* (Coleoptera: Curculionidae)

Distribution

It is a sporadic pest of considerable importance in India. It feeds on the germinating rabi crops.

Damaging symptoms

Adults feed on leaves and tender shoots. Weevils cut the seedlings at the ground level, during October – November, when the rabi crops are germinating.

Life cycle

Adult weevils emerge in June – July and start laying eggs in October – November. A larva takes 20 – 50 days in hatching. Larvae after moulting undergo pupation in 90 days during March – April. The weevil completes one generation in a year.

Management strategy

- Deep summer ploughing helps in reducing the pest population by exposing the pupae to natural enemies like predatory birds.
- Application of Carbaryl 50 WP @ 2 ml/L of water or Malathion 5% dust or Carbofuran 3% CG

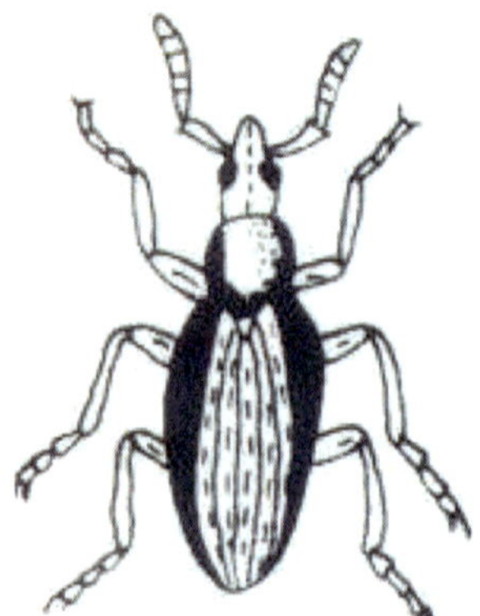

Ghujhia weevil adult

4. Termites, *Odontotermes obesus* Ramb. (Isoptera: Termitidae)

Distribution

The insect is prevalent in all the warmer parts of the world but commonly found in the tropical regions. This is polyphagous pest and feeds on every variety of agricultural crops. Among the crops grown wheat, maize, sorghum, wheat, oat, pea, potato etc.

Nature of damage

This is a polyphagous pest but mainly feeds on wheat and maize. Termites attack both maize stalks and roots. As a result of their attack, the young maize even the shoot of older maize dry up and finally the growth is adversely affected. The maize roots are eaten up from the cut ends and are rendered hollow. In unirrigated areas the young wheat crop is damaged badly, as they destroy the roots of seedlings ad finally the plants are dried off completely. They destroy the crop at night and remain hidden in their nest during day time. The attacked plants can easily be distinguished by irregular cuttings and adhering of soil particles. The fruits trees are also attacked and on branches and stem the soil channels are formed under which they live.

Life history

Three developmental stages viz., egg, nymph and adult are found. They are the social insect and live in colony. There is polymorphism in adult stage and king, queen soldiers, workers are found. The members of the colony may be divided into (i) reproductive forms (queen, king, complimentary form and colonizing form). (ii) Sterile form (workers and soldiers).

The winged reproductive's comes out in swarms from the nest generally in the beginning of monsoon. Swarming usually takes place in the day time and most of the individual of the swarms are destroyed by birds etc. the survivals mate, shed their wings and burrow in the ground to form a new colony of which they become the king and queen. Their flight is known as nuptial flight. The queen lays first batch of 10-130 eggs about a week of swarming, but later on in large numbers about 30,000-40,000 eggs per day throughout her life time, which may be five years. Eggs are small kidney-shaped and yellowish in colour. Egg is 0.5 mm long and incubation period about in a week. The newly hatched nymphs are yellowish white in colour and about 1 mm long. They eat for some time the excreta of the king and queen and latter search the food. Nymphs develop into different castes in 6-13 months after 4-10 moulting.

1. Reproductive form

(a) **Queen:** The queen is the largest individual ranging 6-8 cm in length and 1 cm in thickness. She is mother of the colony and lives in a specially prepared royal chamber which is situated in the centre of the nest at a depth of 1-6 feet below the ground surface. The queen is wingless, creamy white in colour and her abdomen is marked with transverse dark brown bands.

(b) **King:** The winged male which remains with queen after nuptial flight is known as king. There is generally a single king in each colony which is smaller than queen and remains with her in royal chamber.

(c) **Complimentary form:** There are apterous or brachypetrous forms of both the sexes which maintain the numerical strength of the colony in absence of macropterous forms. In a single colony their number may be in hundred.

(d) **Colonizing form:** They are generally produced in rainy season and attracted by light. They are brownish in colour with two pairs of slender, dark brown, long narrowed wings which are used for nuptial flight and after which they are she. These form emerge in millions from the parent colony during monsoon and after finding their mate they craw into the soil and start the new colony. These are thus the kings and the queens of future colonies.

2. Sterile form

(a) **Worker:** The worker constitutes the main labour force of the colony and he is about 80-90% of the whole colony. It is about 6-8 mm long, dirty white in colour with brown head and eyes is small or absent. Mouth part is fairly strong and used as the working tools for all types of odd job.

(b) **Soldier:** The soldier is slightly bigger than worker. Mouth part is well developed and used for very active defense and offence. The main function is to protect the colony but also help the worker is keeping the nest neat and clean by removing the dead and sickly members of the colony. Worker and soldier contain sterile individual of both sexes.

Management

- The removal of dead or decaying matter or dry stubbles from the field useful because they attract the termite.
- The use of partially decomposed manure should be avoided.
- Irrigation water with crude oil emulsion may be used to avoid the termite attack.
- Mixing of insecticide Imidacloprid or Thiamethoxam 30FS at below mentioned rates before sowing in soil has been found very effective.

Cropped area

Wheat : 3 – 4 ml/kg seed
Barley: 4 – 6ml/kg seed
Gram : 15-30ml/kg seed
Soil treatment
Wheat: 2-3 lit./ha.
Sugarcane: 6.25lit/ha

Use 450ml chlorpyriphos 20EC or 600 ml Fipronil 5SL or 500ml Imidacloprid 200SL for seed treatment.

Treat soil with cartap hydrochloride @ 25kg/ha after the last ploughing and before planking in areas where termite attack is recurrent and seed treatment could not be done. For termites control in standing crop, dilute 4L of chlorpyriphos 20EC in 5L of water and mix in 50 kg of sand thoroughly. Broadcast this treated soil in the infested areas.

Never use raw FYM.

Destroy termitaria in and around field.

5. Shoot fly, *Mayetiola destructor* (Say) (Diptera: Cecidomyidae)

Distribution

The fly is capable of infesting and injuring all types of wheat. Although wheat is the preferred host, barley, rye and triticale are other cultivated cereal crops infested occasionally. Several wild grasses also serve as hosts when these crops are not available.

Damaging symptoms

Larva feeds beneath the leaf sheath at the base of the young plants. The specialized mandibles of larva are used to inject salivary fluids into plants. The infested plants show a characteristic stunted appearance with more erect, shorter and darker green leaves than those of uninfested plants. When infestations are severe, stunting of seedlings occurs earlier and many of the young plants die after the larvae have matured. Heavily infested fields show dead plants and thin stands. Feeding causes injury that prevents normal elongation of internodes, thus reducing both quantity and quality of the grain.

Life cycle

Adults do not feed and live for less than 3 days. Adult females lay 100 – 300 eggs between the lengths of wheat leaves. After hatching first-instar larvae crawl down the leaf and settle under the leaves at the joints of stem or at the base of seedlings. They feed on the epidermal cells of developing leaf tissue resulting in gall formation. This causes stunting of plants and also reduces both grain quantity as well as quality.

Management strategy

- Growing shoot fly resistant varieties.
- Cypermethrin 10% EC may be applied.

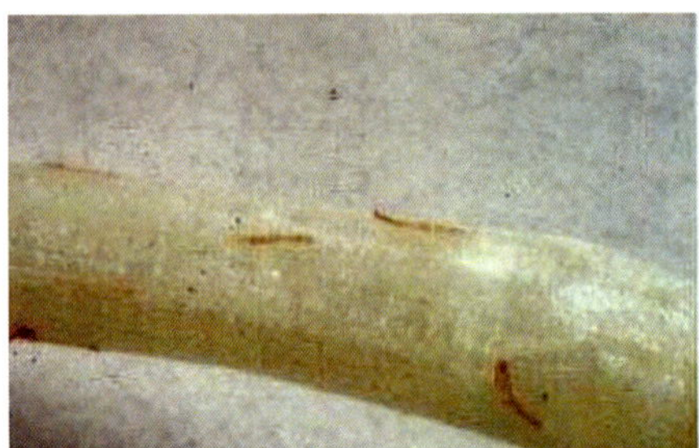
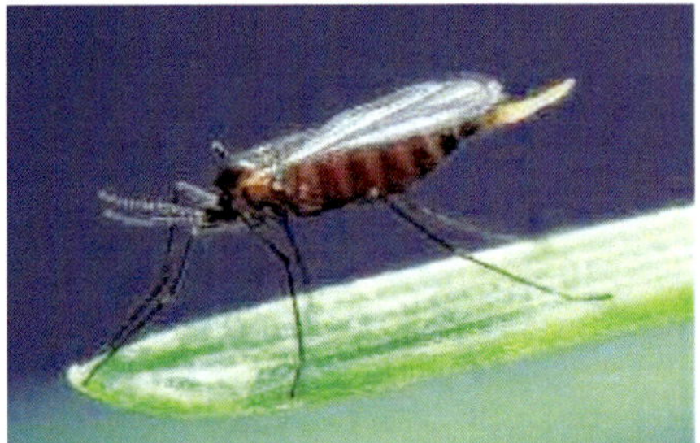

Shoot fly larvae, adult and damaged plants

6. Wheat Gall nematode, *Anguina tritici* (Tylenchida: Tylenchidae)

Damaging symptoms

Affected plants remain stunted and their leaves are wrinkled, rolled or twisted. The diseased ears are shorter and thicker than the healthy ones and the glumes are spread farther apart. Grains in an infested ear produce galls.

The nematode causes ear-cockle or mammi disease throughout the wheat growing parts of the world. These black rounded galls when soaked in water overnight, releases large number of larvae, which looks like wriggling thread-like creatures under microscope. It is also carrier for bacterial yellow slime ear-rot (Tundu disease) caused by *Corynebacterium tritici*. Oats and Barley are however immune to the attack of this nematode.

Life cycle

The dry galls either fall to the ground from ripe ears, or they find their ways to the store. The galls remain dry for long periods and yet the larvae remain viable. A single gall contains 800-30,000 larvae, which are set free into soil, when wheat crop is sown. On reaching the host plant, they rise up the plant and feed as free parasites on young leaves or growing points. Later, they penetrate into the primordial of the flower buds and form galls instead of normal seed. There is only one generation in a year.

Management strategy

- Galls should be separated from wheat seeds by floating them on water. The galls float that can be then easily removed.
- One year fallowing of the infested field is sufficient to eradicate this nematode.

Gall nematodes affected plants and grains

7. Molya nematode, *Heterodera avenae* (Tylenchida: Heteroderidae)

Distribution

The nematode is widely distributed in Europe and Australia and in India it has been recorded from Punjab, Haryana and Rajasthan.

Damaging symptoms

As a result of nematode presence, the main root remains short or bunchy bearing small galls. In case of heavy infestation, seedlings fail to grow, produce short stalks and ears, yielding a poor harvest.

Life cycle

The nematode passes unfavourable season in the form of cysts, mostly in the soil. Cyst consists of dead body of a female, containing large number of eggs, which hatch under favourable conditions, within the cysts. Larvae are then freed into the soil as second infective instars. They invade any underground part of susceptible plant where it moults to become adults. After second and third moultings, female continues to increase in girth, until it becomes ovate. Then it undergoes final / IVth moult to emerge as a full grown female, which dies and become cyst.

Management strategy

- Application of nematicides.

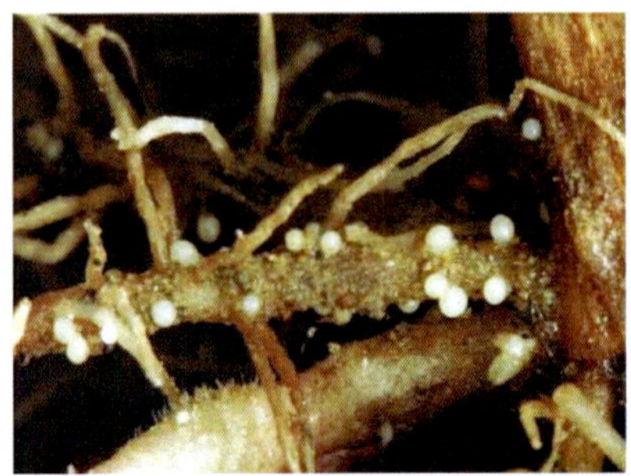

Molya nematode cysts and affected wheat plants

PULSES

- Chickpea
- Lentil
- Field Peas

Chickpea

1. Pod borer, *Helicoverpa armigera* (Hubner) (Lepidoptera: Noctuidae)

Damaging Symptoms

Young larva feeds on tender leaves, buds, flowers and subsequently bores into pods and feeds on the seeds with its head and part of the body only thrust inside, the rest remaining outside.

Life Cycle

Moth is stoutly built, yellowish brown in colour, having a dark speck and a dark area near the outer margin of forewings. Hindwings are whitish and darker in colour. Female lays as many as 741 greenish yellow round eggs singly on tender plant parts. After hatching in 2-4 days, young larvae feed on foliage for some time and later bore into the pods, feeding on developing grains. A single larva may damage as many as 30-40 pods before undergoing pupation in soil.

Management Strategies

- Close spacing or high plant population per unit area attract more pests. Therefore, close planting may be avoided.
- Timely sowing i.e. upto mid October or growing early maturing varieties, so that it escapes the peak pod borer incidence.
- Deep ploughing of the fields in summer reduces the chances of pest propagation.

- Mixed or intercropping with non-preferred hosts like mustard, wheat, coriander, linseed, barley, increases the natural enemy activity and reduces pod borer incidence.
- Installation of pheromone traps @ 5/ha for monitoring and 20-30 /ha for mass trapping of the male moths.
- ETL (Economic threshold limit) is 5-6 moths / trap / day for few days during post winter months.
- Installation of 'T' shaped bird perches @ 50 /ha encourages feeding by insectivorous bird like, black drongo, common myna, etc.
- Conserve larval parasitoid, *Campoletis chloridae* Uchida, as it may parasitize 50-60% of *H. armigera* larvae during vegetative phase and 30-40% during podding stage.
- NPV @ 250 - 500 larval equivalent, L.E. + adjuvants or provides effective control.
- Spray of Methomyl 40 SP @ 1.0 kg/ha to kill eggs.
- Spray of NSKE (1 kg powdered neem kernel soaked in 20 litres of water over night + 200g of ordinary soap) @ 5%.

Gram pod borer larva, pupa and adult

- *Beauveria bassiana* 1% WP, Azadirachtin 0.03% WSP or NPV of *H.a.* 2% AS may be sprayed.
- Carbaryl 10% DP, Chlorpyriphos 1.5% DP, Deltamethrin 2.8% EC, Novaluron 10% EC, Quinalphos 25% EC & 1.5% DP may be applied.

2. Semilooper, *Autographa nigrisigna* Walker (Lepidoptera; Noctuidae)

Damaging Symptoms

Newly hatched larvae scratch chlorophyll from the leaves, while the grown up feed on leaves, buds, flower and pods by nibbling, leaving the basal part of pod with peduncle. As a result of scratching the whole leaf becomes whitish and skeletonized. The pod damage is comparable with the damage caused by *H. armigera* that feeds on grains by making a neat hole on the pods. But this semilooper eats away everything leaving the peduncle only, very similar to the damage caused by birds.

Life Cycle

Egg, larval, pre pupal and pupal stages lasts 3-6, 8-30, 1-3 and 5-13 days respectively. Male moth survives for 4-5 days, while the female lives for 7-9 days. One generation is completed in 18-52 days. This pest is more active during the early vegetative phase of the crop i.e. during November – December. With the advent of cold, its population reduces.

Semilooper larva

Management Strategies

- A tachinid parasitoid has been recorded parasitizing larval and pupal stages of *A. nigrisigna.*
- *Bacillus thuringiensis* formulations are also effective.
- Phosalone (4% dust) @ 25 kg/ ha or malathion or dichlorvos spray gives effective control.

3. Aphids, *Aphis craccivora* Koch (Homoptera: Aphididae)

Damaging Symptoms

Adults as well as nymphs suck the cell sap. As a result of their feeding, there is depletion of assimilates coupled with increased respiration. The leaves and shoots are deformed and stunted followed by sticky honeydew deposition, in case of severe infestation. For early detection inspect new shoots and underside of young leaves.

Life Cycle

Adult aphid is oblong soft bodied insect of about 2mm in length. Apterous females (wingless) are shining dark brown or black, while the alate males (winged) are greenish black in colour. Both winged and wingless forms reproduce viviparously and parthenogenetically laying approximately 1743 numbers within 15 days. Breeding occurs throughout the year on one crop or the other. Besides chickpea, they attack several leguminous crops or weeds such as lentil, mungbean, lathyrus, fenugreek, alfalfa, tomato, potato, etc. Under dry conditions they multiply and spread over larger areas.

Management Strategies

- Wider spacing i.e 60 x 20 cm is more favorable to aphids than narrow spacing i.e. 30 cm.
- Cultivars influence population build up significantly.
- Several coccinellids and syrphids predate on aphids.
- Several insecticides such as acephate dimethoate, fenvalerate and cypermethrin may be used.

Chickpea aphid

4. Hairy caterpillar, *Amsacta moorei* (Butler) (Lepidoptera: Arctiidae)

***Amsacta albistriga* (Walker)**

***Spilosoma obliqua* Walker (Lepidoptera: Arctiidae)**

Damaging Symptoms

Larvae defoliates, thereby affecting grain yield. Young larvae feed on growing points of the plant and on the photosynthetic part from the underside of leaves, while the later instars are voracious feeders and completely skeletonize the crop, in case of heavy infestation. This is a polyphagous insect and feeds practically on all kinds of vegetation during kharif.

Life Cycle

Full grown caterpillars are reddish amber to olive green with their body covered with numerous long hairs, arising from the fleshy tubercles. *A. moorei* moth is stoutly built having white wings with black spots. Outer margin of forewings, anterior margin of thorax and entire abdomen are scarlet red. While the head, thorax and under side of the body are dull yellow in *S. obliqua* moths.

The pest is active during rainy season (Mid –June till end of August) and passes the rest of the year in pupal stage in soil. Moths appear with first shower, and lay yellow spherical eggs in clusters of 700 – 850 each on the under surface of leaves. The young caterpillars after hatching in 2-3 days, feeds on the photosynthetic part of the leaves, thus they become whitish in colour. Such leaves are easily recognizable from a distance and can be plucked along with the congregated larvae and destroyed. Full grown larvae enter the soil and pupate at a depth of about 23 cm. They emerge next year at the onset of monsoon.

Management Strategies

- Deep summer ploughing destroys the pupae.
- The leaves with young congregated larvae can be easily collected and destroyed.
- Collection and destruction of grown-up larvae.
- Moths are strongly attracted to light. Setting up of light traps following the first shower of monsoon and continued throughout the period of emergence for about one month, greatly reduces pest incidence.
- Application of 1.5 kg Carbaryl 50 WP or Quinalphos 25 EC @ 1.25 in 600-1000 L water /ha.

Hairy caterpillar egg mass

First and second instar larvae

Amsacta albistriga

5. Cut worm, *Agrotis ipsilon* (Hufnagel) (Lepidoptera: Noctuidae)

Agrotis flammatra Schiff. (Lepidoptera: Noctuidae)

Damaging Symptoms

Larva become active at night and damages the seedlings by cutting at the base or just above soil surface. The cut plants or twigs are dragged and partially buried into the soil. This indicates their presence in the field. During day, the caterpillars remain hidden in cracks or within clods.

Life Cycle

A single female moth can lay as many as 639 – 2252 eggs on earthen clods, chickpea stem bases or on both sides of leaves. The average incubation, larval, pupal and total duration varies from 2.7 – 5.1, 18.2 – 39.5, 10.3 – 25.3 and 31.4 – 69.8 days respectively. Extremely dry weather during April – May adversely affects cutworm multiplication. During off-season (summer and rainy), the pest remains scattered among weeds growing nearby.

Management Strategies

- Field ploughing 4-5 times before sowing rabi crops breaks the clods in the field and reduces pest incidence.
- Clean cultivation, i.e. cleaning the weeds from bunds, nearby fields.
- Braconids, *Microgaster* sp., *Bracon kitcheneri* and *Fileanta ruficanda* parasitizes *Agrotis* larvae, while *Broscus punctatus* and *Liogrullus bimaculatus* predate on cutworms.
- Mix 2L of chlorpyriphos 20 EC in 25 kg of sand/ha in rows of the plants at sowing time.

Damage caused by cut worm larva

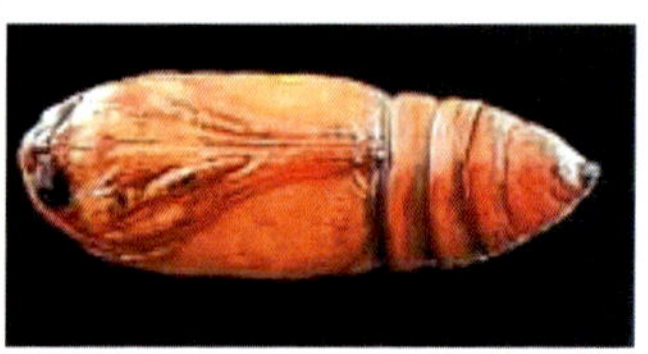

Cut worm larva, pupa and adult

Field pea

1. Pea leaf miner, *Chromatomyia horticola* (Goureau) (Diptera: Agromyzidae)

Damaging Symptoms

Larvae damages by making serpentine mines in the leaves. Usually mine starts from the periphery and ends towards the mid rib. As larva grows, the mine widens and becomes distinct marked by dark excretal pellets. As a result of their feeding, photosynthesis is reduced and growth of the plant is hampered. Adults with their sucking and sponging type of mouth parts feed on cell sap. Adults also feed on pollen. Flowering and fruiting of the infected crop is much reduced.

Life Cycle

Eggs are laid singly in leaf tissues in the beginning of December. Larvae after hatching in 2-3 days feed between upper and lower epidermis by making zig-zag tunnels, leaving behind a trail of faecal matter inside the mines. They become full grown in about 5 days and pupate within the galleries itself. Adults emerge in 6 days and the total life cycle is complete in 13-14 days. Adult flies are greyish with black dorsal setae. There are several generations in a year.

Management Strategies

- Spraying with Dimethoate 0.0%, Phosphamidon 0.1% or 0.05% malathion 50 EC when attack starts effectively checks the pest damage.
- Spraying may be repeated after 15 days, if necessary.
- Spray 1.0 L of dimethoate 30 EC in 750 L of water per ha and repeat spray at I 5 day interval. A waiting period of 20 days should be observed before harvest.

Pea leaf miner larvae and pupae forming galleries on leaves

Pea leaf miner attack on pods

Adult

2. Pea pod borer, *Etiella zinckenella* Treitschke (Lepidoptera: Phycitidae)

Damaging Symptoms

Newly emerged tiny greenish caterpillars feed on floral parts and subsequently bore into the pods to feed on the seeds. Dropping of flowers and young pods is noticed as a result of their feeding, while the older pods are marked with a brown spot from where the larva enters. Up to 5% reduction in yield may be noticed as a result of their feeding. It is a serious pest of lentil and green pea in northern India and it also attacks variety of other pulses in various parts of the country, Myanmar and Sri Lanka.

Life Cycle

Eggs are laid either singly or in clusters on various parts of plant preferably pods at reproductive stage. Larvae on hatching in 5 days, start feeding on floral parts and later bore into the pods to feed on seeds. Larvae after maturing in 10-27 days, pupate in soil at a depth of 2-4 cm and the pupal development is complete in 10-15 days. Young larvae are greenish, which later turn rosy, with a purplish tinge. Moths are grey with a wing span of 25 mm. Forewings have dark marginal lines and the prothorax is orange coloured. It completes 5 generations in a year and breeds throughout the year.

Management Strategies

- Spraying with Carbaryl 0.05% at flower initiation stage effectively controls the pest for at least three weeks.
- Spraying with Flubendiamide 39.35 EC @ 75 ml/ha or Indoxacarb 14.5 SC @ 500ml/ha at flower initiation and repeated spray after 15 days of first spray if necessary, effectively manages this pest.

Pea pod borer larva and adult

3. Gram pod borer Helicoverpa armigera (Hubner) (Lepidoptera: Noctuidae)

Damaging symptoms, life cycle and management strategies

- Same as for insect pests of chickpea.

4. Pea stem fly, *Melanogromyza phaseoli* Tryon (Diptera: Agromyzidae)

Damaging Symptoms

Creamy maggot damages by tunneling through the stem of young seedlings and later on the growing plants. Drooping of the tender leaves and wilting of seedlings is noticed when attacked at the early growing period. In case of severe infestation, leaves turn yellow giving a dry appearance to plants. Stem turn brown, becomes swollen and breakdown, where maggot and pupae are present. Attacked plants bear few pods, which are mostly empty or bear very small seeds.

Life Cycle

Maggots are small yellow coloured, while the adults are tiny black flies, 5 mm in length. Eggs are laid on the wall of the immature pod. The maggot on hatching feed on developing seed and grows to a length of 3 mm before pupating.

Management Strategies

- Avoid early sowing. Sow the crop in 2nd fortnight of October to escape

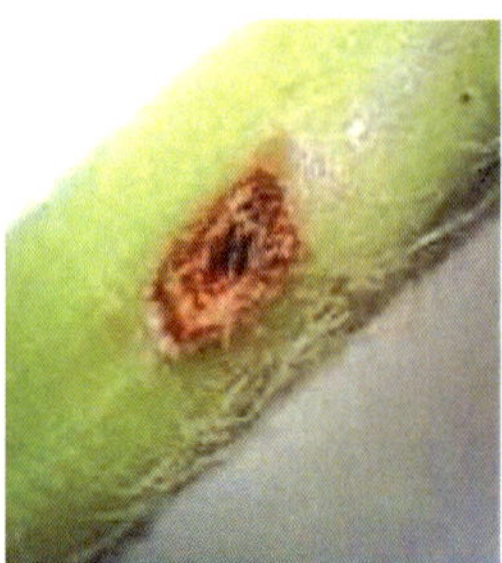

Damage caused by pea stem fly, their maggot and adult fly

damage by this pest.

- Seed treatment with chlorpyriphos may be done before sowing.
- Soil application of 10 kg phorate 10G is effective up to 40 days of sowing.

5. Pea blue butterfly

***Lampides boeticus* Linnaeus**

***Euchrysops (Catochrysops) cnejus* (Lepidoptera: Lycanidae)**

Damaging Symptoms

Larvae starts feeding on tender leaves and later bore into the buds, flowers, green pods and feed inside the pod on the developing grains.

Life Cycle

E. cnejus larvae are flat and slightly rounded and hairy. Adults have black spots on the ventral sides of their wings. *L. boeticus* larvae are pale green in colour, while the adult has numerous stripes on the ventral side of wings.

The male is dull purple above, with two black tornal spots on the hindwing. The female is brown, with the wing bases pale shining blue. The underside is buff, transversed by white fasciae. The black tornal spots are orange-crowned. The white tipped tail is about 3-5 mm long.

Pea blue butterfly adult

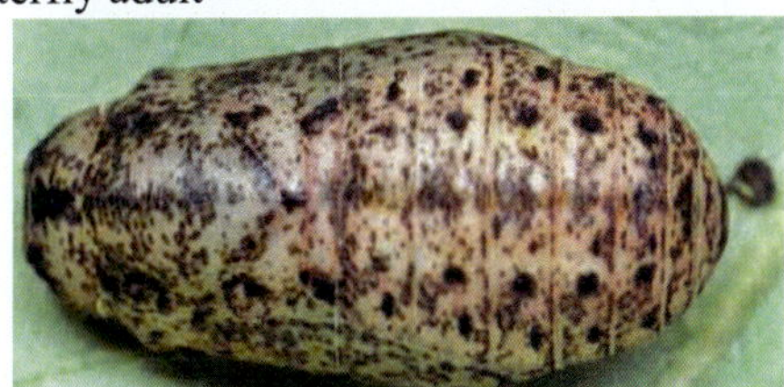

Blue butterfly larva and pupa

Management Strategies

- Same as for pea pod borer.

6. Pea green aphid, *Acyrthosiphon pisum* (Harris) S. lat. (Hemiptera: Aphididae)

Distribution

It is oligophagous specie, feeds on only some species of legumes, while the other species, like green peach aphid, *Myzus persicae* (Sulzer) is a polyphagous specie, which feeds on over 400 plant species.

Damaging Symptoms

Both adults and nymphs suck cell sap mostly from growing parts of the plants, thereby retarding development and sometimes causing complete loss of crop. Plants become stunted and yellow. Leaves and pods are malformed and grains undersized.

Life Cycle

Management Strategies

- Spraying with phosphamidon or dimethoate when attack starts effectively manages the aphid population.
- Spraying may be repeated after 15 days, if necessary.

Green pea aphid nymphs on leaves and pods

Pea aphid adult

Aphid parasitized by a parasitoid wasp

7. Cut worm

Damaging symptoms, life cycle and management strategies

- Same as for insect pests of chickpea.

Lentil

1. Lentil pod borer, *Etiella zinckenella* (Lepidoptera: Phycitidae)

Damaging symptoms, life cycle and management strategies

- Same as for insect pests of field pea.

2. Black aphids, *Aphis craccivora* Koch (Homoptera: Aphididae)

Damaging symptoms, life cycle and management strategies

- Same as for insect pests of chickpea.

3. Cut worm, *Agrotis ipsilon* (Hufnagel) (Lepidoptera: Noctuidae)

Damaging symptoms, life cycle and management strategies

- Same as for insect pests of chickpea.

Lathyrus

1. Pod borer, *Helicoverpa armigera*

Damaging symptoms, life cycle and management strategies

- Same as for insect pests of chickpea.

2. Aphids, *Aphis craccivora* Koch (Homoptera: Aphididae)

Damaging symptoms, life cycle and management strategies

- Same as for insect pests of chickpea.

3. Bean bugs, *Clavigralla gibbosa* Spinola (Hemiptera: Coreidae)

Dusky cotton bug, *Oxycarenus laetus* Kirby (Hemiptera: Lygaeidae)

Damaging Symptoms

Both nymphs and adults suck cell sap from leaves, stem, pods, flower buds. Pods show yellow patches and later pods as well as grains shrivel in case of heavy attack. The grain size reduces thereby reducing the yield significantly.

Life Cycle

C. gibbosa

Adult bugs are greenish brown in colour, 20 mm long with spined pronotum and swollen femur. A single female lays as many as 52 eggs usually on pods and sometimes on leaves and other floral buds. The mean egg, nymphal and adult periods of *C. gibbosa* on field bean is 8, 20 and 11 days. The total life cycle of *C. gibbosa* is 42 days. The pest is active from middle of October till end of May and has six overlapping generations during this period.

O. laetus

The female bug lays transparent, light yellow, oval or cigar shaped eggs either singly or in groups of 2 - 10 usually on pods. The freshly hatched light pink coloured nymphs hatch in 8 days with a small black tail like process. The total nymphal period ranges from 15 to 23 days. Freshly emerged adult is bright red in colour. Body colour of adult turns to dark brown after four to five hours of moulting. The adult bugs are small, flat with pointed head and uniform dark brown body with dirty white or dusky brown transparent hemielytra. Total life cycle of varies from 35 to 50 days.

Management Strategies

- Shaking the infested plants over vessels of oil and water or oily cloth gives very effective control.
- Egg masses of bugs should be collected and kept away from the cropped area as 30% of the egg masses are parasitized by *Gryon* sp. This practice will allow the parasite to be released in nature but will prevent the nymphs from causing damage.

Bean bug adult damaging pods

Horse Gram

1. Pod borer, *E. zinckenella* (Lepidoptera:Phycitidae)

Damaging symptoms, life cycle and management strategies

- Same as for insect pests of field pea.

2. Hairy caterpillar

Damaging symptoms, life cycle and management strategies

- Same as for insect pests of chickpea.

3. Aphids, *Aphis craccivora*

Damaging symptoms, life cycle and management strategies

- Same as for insect pests of chickpea.

4. Whitefly and Green Jassid

Damaging Symptoms

Both adults and nymphs suck the cell sap, thus devitalizing the plant and are also vector of Yellow Mosaic Virus (YMV). They also excrete honeydew on which the sooty mould grows, interfering with the normal photosynthesis process. Consequently the crop gives sickly, black appearance from a distance. It transmits a number of viral diseases in several crops.

Life Cycle

Females lay approximately 119 eggs singly on the underside of the leaves. The elliptical, stalked, yellowish coloured eggs hatch in 3-5 days in April – September, 5-17 days in October – November and 33 days December – January. The nymphs suck cell sap and moults thrice to form pupae within 9-14 days in April – September and 17-81 days during winter months. The total life cycle is completed in 14 – 122 days and there are 11 generations in a year.

The nymphs are louse-like sluggish creature, while the adult are pale yellow, 1.0-1.5 mm long, with white waxy powdery covering.

Management Strategies

- Early sowing during spring (15th March) suffers less damage.
- Inter/mixed cropping with taller crops shields the crop, thereby checking the dispersal of flying insects.

Whitefly adults

Whitefly adults sticking to the Yellow sticky trap

- Spraying the crop at bud initiation stage with Dimethoate, Malathion, Metasystox or Phosphamidon effectively controls this pest.

5. Plume moth, *Exelastis atomosa* (Lepidoptera: Pterophoridae)

Damaging symptoms

Young larvae bore into the unopened flower buds for consuming the developing anthers. Grown up larvae bore into pods, however it never enters the pod completely.

Life cycle

Management strategies

- Spraying with Carbaryl 0.05% at flower initiation stage effectively controls the pest for at least three weeks.

Plume moth larva and adult

Oilseeds

Mustard / Toria / Gobhi Sarson

1. Mustard Aphid (*Lipaphis erysismi*); Hemiptera : Aphididae

Damaging symptoms

Both nymphs and adults suck the sap from leaves, buds and pods. Curling may occur in infested leaves and at advanced stage plants may wither and die. Plants remain stunted and sooty molds grow on the honey dew excreted by the insects. The infected field looks sickly and blighted in appearance.

Life cycle

Aphids are small, soft-bodied, pearl-shaped insects that have a pair of cornicles (wax-secreting tubes) projecting out from the fifth or sixth abdominal segment. They attacks generally during 2nd and 3rd week of December and continues till March. Rainy and humid weather help in accelerating the growth of insects.

Management strategies

- Sow the crop before 20th October to escape the peak aphid incidence.
- Install yellow stick trap to monitor aphid population.
- Destroy the affected parts along with aphid population in the initial stage.
- Conservation of the natural enemies: Ladybird beetles viz., *Coccinella septempunctata, Menochilus sexmaculata, Hippodamia variegata and Cheilomones vicina* are most efficient pradators of the mustard aphid. Adult beetles may feed on an average of 10 to 15 adults /day.
- Several species of syrphid fly i.e., *Sphaerophoria* spp., *Eristallis* spp., *Metasyrphis* spp., *Xanthogramma* spp and *Syrphus* spp. are predating on aphids.
- The braconid parasitoid, *Diaretiella rapae* a very active bio control agent cause the mummification of aphids.
- The lacewing, *Chrysoperla carnea* predates on the mustard aphid colony.
- A number of entomogenous fungi, *Cephalosporium* spp., *Entomophthora* and *Verticillium lecanii* infect aphids.
- Apply insecticides when the economic threshold level of 40- 45% plant infestation or 50-60 aphids per 10 cm central shoot are observed.
- Spray the crop with one of the following in the flowering stage; Chlorpyriphos 20% EC Dimethoate 30% EC, Malathion 50% EC, Methyl Parathion 2% DP, Oxydemeton methyl 25% EC, Phorate 10% CG, Phosphamidon 40% SL Thiamethoxom 25% WG @ 625-1000 ml per ha.
- Combination of chlorpyriphos +acetamiprid 0.05% is highly effective against mustard aphid.

Lady bird beetle pupa and adult

Parasitized mustard aphid mummies

2. Mustard saw fly: *Athalia lugens proxima* (Hymenoptera: Tenthredinidae)

Damaging Symptoms

Initially the larva nibbles leaves, later it feeds from the margins towards the midrib. The grubs cause numerous shot holes and even riddled the entire leaves by voracious feeding. They devour the epidermis of the shoot, resulting in drying up of seedlings and failure to bear seeds in older plants.

Life cycle

Larva are greenish black with wrinkled body and has eight pairs of pro-legs. On touch the larva falls to ground and feigns death. Adult head and thorax is black in colour while abdomen is orange coloured. Wings are translucent, smoky with black veins.

Management strategies

- Deep summer ploughing to destroy the pupa.
- Early sowing should be done.
- Maintain clean cultivation.
- Severe cold reduces pest load.
- Collection and destruction of grubs of saw fly in morning and evening
- Use of bitter gourd seed oil emulsion as on antifeedant.
- Spray the crop with malathion 50 EC @ 1000 ml/ha quinalphos 25 EC @ 625ml/ha. All this should be applied in about 600 to 700 litres of water per ha.

3. Leaf webber: *Crocidolomia binotalis* (Lepidoptera: Pyralidae)

Damaging Symptoms

Newly hatched larvae feed initially on the chlorophyll of young leaves and later on older leaves, buds and pods, make webbings and live within. Severely attacked plants are defoliated. Seeds in the pods are also eaten away.

Life cycle

Adult moths are yellowish-brown, with reddish-brown forewings and distinct wavy lines and prominent white spots, while the hind wings are white with dark brown apical area. Eggs hatch in 5 – 15 days. Larval period varies from 24 – 27 days. Larvae are pale yellowish-brown, with a series of lateral and sub-lateral black spots and specks. Pupal period varies from 14 – 40 days and the total life cycle is completed in about 45 days during summer.

Management

- NSKE 5% or *B.t.* @ 2g/l water may be applied.
- Phosalone 35 EC @ 2 ml/ l, fenvalerate 20 EC 0.5 ml/lit or Carbaryl 10% D.P. may be sprayed.

Leaf webber larvae and adult

4. Leaf miner: *Chromatomyia horticola (Phytomyza atricornis)* (Diptera: Agromyzidae)

Damaging Symptoms

Young maggot mines zig-zag galleries in the leaves, thus reducing the photosynthetic leaf area.

Life cycle

Larvae are small whitish maggot, which pupates inside the mine itself and only the adult comes out of the mine.

Management

Spray 5 % neem seed kernel extract (NSKE)

Leaf miner maggot and adult

5. Cabbage head borer: *Hellula undalis* (Lepidoptera: Crambidae)

Damaging Symptoms

Caterpillars initially mine the leaves and make it white papery. Later they feed on leaves and bore into stems. Entrance hole is covered with silk and excreta.

Life cycle

Larva is pale whitish brown with 4 or 5 pinkish-brown longitudinal stripes. Adult moths are pale greyish-brown, suffused with reddish colour. Forewings have grey wavy lines, an apical spot and pale edged dark lunule. Hind wings are pale dusky, darker in apical area.

Management strategies

- Collection and careful destruction of the larvae at gregarious stage with leaves twice a week.
- For control of grown up larvae apply 5% malathion dust @ 37.5 kg/ha or carbaryl 50 WP @ 375 g in 150 litre of water.

Cabbage head borer larva, pupa and adult

6. Diamondback moth: *Plutella xylostella* (Lepidoptera: Plutellidae)

Damaging Symptoms

Whitish patches due to scrapping of epidermal leaf tissues by young larvae. The leaves give a withered appearance but in later stages larvae bore holes in the leaves. Leaves may be eaten up completely. It also bores into pods and feeds developing seed.

Life cycle

Diamondback moth larvae are yellowish green, with fine erect black hairs scattered all over the body. Adult moth are small greyish brown having pale whitish narrow wings with yellow inner margin. Forewings have three white triangular spots along the inner-margin, which appear as diamond shaped, hence the name diamondback moth. Hind wings – have a fringe of long fine hair. Entire life cycle is completed in 14 – 50 days.

Management strategies

- Installing pheromone trap @ 5/ acre to monitor the moth activity, and @ 20 /ha for mass trapping of the male moths.
- Collection and careful destruction of the larvae at gregarious stage at least twice a week.

Pods damaged by Diamondback moth

DBM larva and pupa

Diamondback adult

- Conserve *Cotesia plutellae* and *Diadegma insulare* as it is an important parasitoid for diamond back moth.
- For control of grown up larvae apply 5% Malathion dust @ 37.5 kg/ha.

7. Painted bug: *Bargrada hilaris cruciferarum* (Hemiptera: Pentatomidae)

Damaging symptoms

Both adults and nymphs suck the sap from pods mainly. Adult bugs also excrete resinous substances which spoils the pods. Young plants wilt and wither as a result of the attack.

Life cycle

Adult bug is black in colour with red and yellow lines.

Management strategies

- Deep ploughing so that the eggs of painted bugs are destroyed.
- Early sowing is needed to avoid pest attack.
- Irrigate the crop during four week after sowing to reduce pest attack
- Quick threshing of the harvested crop should be done.
- Burn the remains of mustard crop so that the stages of insect do not reach the next year crop.
- The bugs usually congregate on the leaves and stem which can be jerked to dislodge them and killed in kerosinised water.
- Conserve bio-control agents like *Alophara spp*.
- Spray the crop with malathion 50 EC @1000 ml or dimethoate 30EC @ 625 ml in 600- 700 liter water

Painted Bug Adult

Polyphagous Pests

1. Rodents

Rodents are one of the most important non - insects pests of agricultural crops, particularly *kharif* and *rabi* cereal.

Types of rodents

- Lesser bandicoot rat: *Bandicota bengalensis*
- Field mouse: *Mus booduga*
- Indian gerbil: *Tatera Indica*
- Soft furred field rat : *Rattus meltada*

Damage at different stages: In India, rodents have been estimated to cause 5 to 10% losses in cereal crops. Among the field crops, rice and wheat are the most vulnerable crop to rodents. In addition to tiller cutting, they also hoard ripened panicles inside their burrows.

Nursery in rice: The nurseries are drained out and the rodents run freely inside the bed spoiling all germinated seed. Later, they also cut the seedlings 1-2 inches above the water level.

Main field: Sometimes the rodents pull out the transplanted seedlings in paddy and create gaps in the main field. Generally, their activity is confined to inside field leaving 2-4 meters on all sides of the field. In the initial stage, damage appears in patches and after some time, all these small patches become into one big patch. Damage increases with the onset of panicle initiation and continues up to panicle emergence.

Management of Rodents

- In Local traps called 'butta' are extensively used for the control of rodents in cereal. These traps provide fairly good results when applied after chemical control operation.
- However when directly used, trapping will be costly affair and one cannot manage entire population over large areas.
- Moreover, at certain crop stages, like primordial formation, rodents are not attracted towards traps.

Natural Smoke

- The main principle involved in this operation is simply filling the burrows with smoke, which causes suffocation to rodents ultimately leading to their death.
- The smoke liberated by burning rice straw mainly contains carbondioxide.

Chemical Control Fumigation

- The fumigants like Aluminum phosphide is effective and widely used for the control of field rodents living in burrows.
- The control of rodents using rodenticides is the more common way

1. Acute rodenticides (Single dose and quick acting), Eg : Zinc phosphide.
2. Chronic rodenticides (Multi dose and slow acting), Eg : Warfarin, Bromodiolone.

1) Acute Rodenticides: Among the acute rodenticides, Zinc phosphide and Barium carbonate are registered for use.

Action plan for Rodent control

Day 1 Identify live burrows and place 20 g of pre-bait Material inside the burrow.

Day 3 Place 10 g Zinc phosphide poison bait inside the Burrow.

Day 4 Collect dead rats and bury them. Close all the Burrows.

Day 5 Eliminate the residual population through trapping or Burrow fumigation with burrow fumigator. Treat the opened burrows with aluminum phosphide 2 pellets per burrow.

Day 13 In dry black soils fumigation will not give results. Hence apply Bromodiolone 1 cake per burrow. Repeat Bromodiolone baiting.

Advantages of Zinc Phosphide

1. Quick killing
2. Small quantity of chemical is required
3. Single feeding
4. Population can be brought down immediately.

Disadvantages of Zinc Phosphide

1. Necessity of prebaiting
2. Low killing around 40 - 50 %
3. Induce bait shyness
4. Toxic to non target species
5. Chances of secondary poisoning are more

2) Chronic Rodenticides

- In order to overcome limitations and hazardous nature of acute rodenticides, lengthy baiting programme and possibility of resistance, new series of rodenticides have been developed and known as single dose anti-coagulants or second generation anti-coagulants. These rodenticides (Bromodiolone) combines better qualities of acute and chronic rodenticides.
- For effective and successful rodent control, the following programme should be adopted on large areas at a time on community approach.

- Chronic or multi-dose rodenticides (at present only anti-coagulants) are much safer than acute rodenticides because they are less toxic to the non-target species. But they have to be fed for 5 - 7 days to obtain desired results, which increases the cost of operation.

Day-1: Identify live burrows and place 15 g bromodiolone concentrate bait inside the burrow.

Day-2: Repeat Bromodiolone baiting in active or live burrows.

Day-3: Eliminate residual population through trapping or fumigation with burrow fumigator.

Principles

- Grow same maturity group cultivars on large areas to restrict the availability of the vulnerable stage (Reproductive) of the crop.
- Reduce the number and size of the bunds, keep them clean to locate burrows and avoid harborage.
- Rodent control operations should be taken up on large area at a time.
- It checks cross infestation or migration of rodents from untreated fields to treated fields.
- All the control operations should be completed before the crop attains primordial initiation stage since at this stage the rodents are invariably attracted to the rice crop.
- Rodenticides should be made available before beginning of the season.

Section - B
Major Diseases of Rainfed Rabi Crops

2

Cereals

Wheat

1. Yellow rust or stripe rust

Causal organism- ***Puccinia striiformis*** **West**

Nature and recurrence of disease: The disease is air borne. The inoculum (uredospores) causing the annual recurrence is brought from the hills to the plains every year. The uredospores of *Puccinia striiformis* over-summer on the hills at 7000 feet and above.

Symptoms

- The pustules are formed on the leaves but in sever attacks they are formed on the leaf sheaths, stalks and glumes also.
- The green colour of the leaves fades in long streaks on which rows of pustules appear. The pustules are oval and lemon in shape, yellow in colour is arranged in rows.
- The arrangement of lemon yellow shaped pustules in rows is the characteristic feature of yellow rust.
- In sever cases these rows are not district as the large patches of leaf become covered with crowded pustules and the ground of wheat field may appear orange due to a large number of spores falling on it.

Management

- Use resistant varieties and adopt timely sowing.
- Apply balance dose of nitrogen & potash.
- Spray the crop with Propiconazole @ 0.1% or Mancozeb @ 0.25%. Ist spray is to be given at the initiation of disease and repeat spraying after 10-14 days intervals.

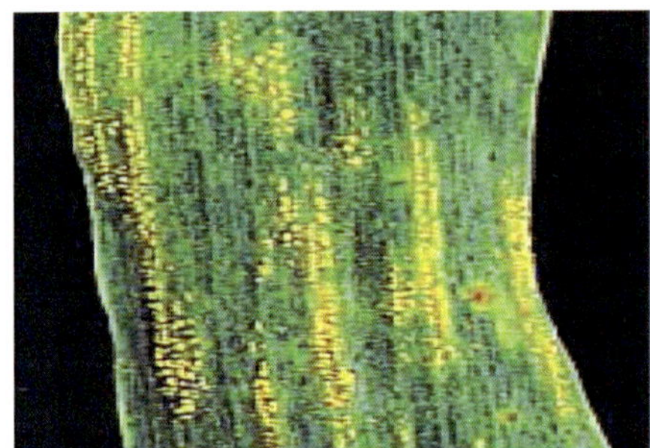

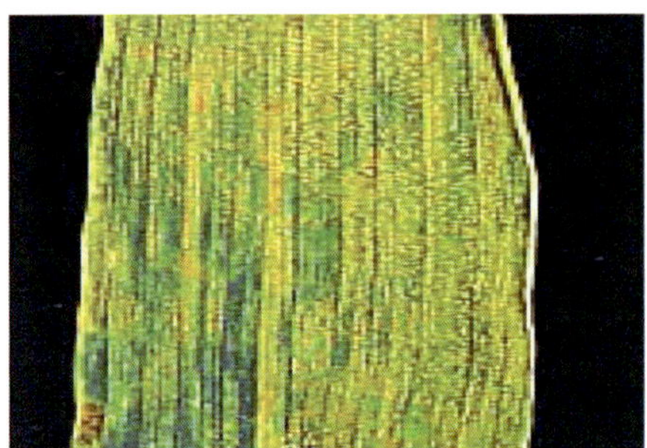

Yellow rust

2. Brown rust

Causal organism: *Puccinia recondita* Rob.ex.Desm

Nature and recurrence of disease: The disease is air-borne. Like other rusts of wheat the inoculum (uredosproes) survives on the tillers and self sown plants in the hills. The inoculum survives at 5,000 feet and above. Every year the uredospores are carried over by wind from the hills to the plains, where they cause infection.

Symptoms

- This is the earliest rust to appear as small, round, light yellowish brown pustules on the leaves.
- The pustules mostly develops on the leaf and leaf sheaths and are never in rows but are either in small cluster or irregular scattered.

Management

- Use resistant varieties and adopt timely sowing.
- Apply balance dose of nitrogen & potash.
- Spray the crop with Tilt 25EC / Shine 25EC/Bumper25EC or Folicur 25EC @200ml or Bayleton 25WP @200g in 200litres of water/acre as soon as the disease is noticed. Repeat the spray at 15 days interval.

Brown rust of wheat

3. Loose smut

Causal organism: *Ustilago tritici* (Pers) Rostr.

Nature and Recurrence: The disease is seed- borne (internally seed borne). The dormant mycelia penetrate inside the embryos of infected seeds and whenever such seeds are sown the seedlings become infected from the very beginning. The disease happens to be internally seed borne and systemic.

Symptoms

- The ear heads are smutted in which the grains replaced by a black powdery mass of spores and finally only the naked rachis is remaining behind.
- Almost every ear of a diseased plant is converted into black loose powdery mass of spores.

Management

- Use resistant varieties.
- Rogue out the affected ear heads of yellowing of flag leaves.
- **Solar heat treatment:** Soak the wheat seed in ordinary water from 8a.m. to 12 noon on any calm and sunny day during May/June. After 4hours soaking, spread out the moist seed in the sun in a thin layer on cemented floor (pucca), on tarpaulin or sheets of cloth. Dry the grain completely and store in a dry place till sowing.
- Treat the seed with Captan or Thiram @2g/kg.

Loose smut of wheat

4. Flag smut

Causal organism: *Urocystis agropyri* (Preuss) Schroet.

Nature and recurrence of disease: The disease is seed-borne and soil borne. Infection occurs below ground. The spore balls are adapted to retain their power of germination for a long time. On germination they infect the primary shoot. The fungus is systemic and later produces the characteristic spores bearing sori.

Symptoms

- The smut lesions appear as long, light green stripes which soon become grayish black streaks parallel to veins are formed on leaves.
- The streaks between veins cause leaves to roll and twist. In advance stage of growth the stripes eventually rupture and expose black sooty masses of spores.

Management

- Practise shallow sowing.
- Rogue out the affected stools and destroy them by burning.
- Dress the wheat seed with Vitavax Power @ 2g or Vitavax @2g or Raxil @1g per kg seed or Bavistin or Agrozim or Derosal or Sten 50 or Benlate @2.5g or Thiram 75%@2g/kg seed before sowing.

Flag smut of wheat

5. Karnal bunt

Causal organism: *Noevossia indica*

Nature and recurrence of disease: The disease is soil-borne as well as air-borne. The spores found in the soil germinate, forming a large number of needle-shaped sporidia (basidiospores), and sickle-shaped secondary sporidia. The sporidia are being carried to the inflorescence through the air currents, where they infect the ovaries and the developing kernals. Infection takes place only when the flowering occurs.

Symptoms

- The disease appears when the grains have developed. In this disease some grains in the spike are partially or wholly converted into black powdery masses, which smells like rotten fish only few ears and few grains in a ears are affected.

Management

- Seed treatment same as flag smut will partially check karnal bunt.
- Use resistant varieties.
- Use disease free seeds.
- Spray the crop with Tilt/Folicur 25EC @200ml/acre using 200litres at ear emergence stage for the control of Karnal bunt is recommended in wheat meant for seed production only.

Karnal bunt of wheat

6. Powdery mildew

Causal organism: *Erysiphe graminis* f. sp. *tritici* E. March

Nature and recurrence of disease: The recurrence of the disease takes place through the cleistothecia (the perenating bodies). The cleoistothecia perenate on the straw and plant debris after harvest, and provide the necessary inoculum (ascospores) for the next season. The primary infection is brought by ascospores, and the secondary infection is from the air-borne conidia.

Symptoms

- Typical light grey to white powdery mass appears on all areal parts of plant.

Management

- Grow resistant varieties.
- Spray the crop with Dinocap @ 0.05% - 0.1% or Wettable sulphur @ 0.25% or Calaxin 0.1% if incidence is noticed.
- Spray with Karathane 40EC @ 0.05% (100ml in 200litres of water) or with Tilt 25EC 200ml in 200litres of water per acre.

Powdery mildew of wheat

7. Leaf blight

Causal organism: *Alternaria / Drechslera spp.*

Nature and recurrence of disease: This is a seed as well as soil-borne disease. The intensity of the disease varies of different varieties and on an average it is 12.2 percent. The inoculum is found to be seated in the seed. Seven to fifteen days old seedlings are not susceptible. As the plants become older, they become more susceptible to the disease. If the suitable conditions of moisture and temperature prevail, the ten week old plants are heavily infected by the disease.

Symptoms

- The initial symptoms appears as small, oval , discolored radish brown spots, irregularly scattered on the lower most leaves.
- The spots are surrounded by bright yellow marginal zone.
- Later on the spots enlarge and coalesce and cover the larger areas giving the leaf a blighted appearance.

Management

- Grow resistant varieties.
- On the appearance of disease spray the crop with Zineb or Mancozeb @ 0.25% or Propiconazole @ 0.1%.
- Treat the seed with 3g Thiram or Captaf (83%) per kg of seed.

Leaf blight

Barley

1. Leaf Stripe

Casual organism: *Drechslera graminium*

Symptoms

- Yellow to brown stripes on leaves
- Plants become stunted with sledded leaves and wither away due to systemic infection

Management

- Before sowing treat the seed with Vitavax or Raxil @1.5g/kg of seed to control leaf stripe

Leaf stripe

- Solar heat treatment as recommended for wheat can be done to control the loose.
- Use resistant variety

2. Covered smut

Casual organism: *Ustilago hordei*

Nature and Recurrence of disease: This is an externally seed-borne and systemic disease. Every year the recurrence of disease takes place by the contaminated seeds. The mycelium develops from the very beginning, along with the coleoptiles and ultimately reaches the ear. During threshing the smut sori break, the smut spores are released and they make the seeds contaminated again.

Symptoms

- Except the awns, the entire ear turns into black compact mass of spores.

Management

- Late and shallow sowing to reduce seedling infection.
- Before sowing treat the seed with Vitavax or Raxil @1.5g/kg of seed to control leaf stripe
- Solar heat treatment as recommended for wheat can be done to control the loose.
- Use resistant variety.

Covered smut of barley

3. Loose Smut

Casual organism: *Ustilago nuda*

Nature and Recurrence of disease: This is a soil and seed borne disease. The smut spores remain scattered over the grains at harvest and threshing time and carried to the store houses. If not disinfected these grains are sources of infection at planting time. Other spores are scattered over the soil, or old refuse on the soil, and thus are a source of infection for the next year.

Symptoms

- Disease that can destroy a large proportion of a barley crop. Loose smut replaces grain heads with smut, or masses of spores which infect the open flowers of healthy plants and grow into the seed, without showing any symptoms.
- The smut sori are enclosed in a fragile membrane which soon ruptures, resealing the dark spore mass which is disseminated by wind, leaves the naked rachis behind.

Management

- The smut sori are enclosed in a fragile membrane which soon ruptures, resealing the dark spore mass which is disseminated by wind, leaves the naked rachis behind. Before sowing treat the seed with Vitavax or Raxil @1.5g per kg of seed to control leaf stripe. Solar heat treatment as recommended for wheat can be done to control the loose.

Loose smut of wheat

Pulses

Chickpea

1. Ascochyta blight

Causal organism: *Ascochyta rabiei.*

Nature and recurrence of disease: The disease may survive in the soil along with plant debris. The infection may take place by means of conidia found in the soil:

Symptoms

- The symptoms appear as dark brown spots studded with dot like bodies are produced on the stem, branches, leaf lets, and pods.
- All the green parts of the plant attacked. Dark lesions appear on the stem and leaves, then on the pods.
- On the stem, the lesions are oval or elongated, whereas on the leaves and pods, round and upto 1.25cm in diameter.
- The disease spreads rapidly due to favourable climatic conditions like rain and humid weather. In severe infection whole plant may be blighted and killed.

Management

- Grow resistant varieties.
- Follow crop rotation and crop sanitation.
- Treat the seeds with Carbendazim (50WP) Thiram @ 3g/kg seed or carbendazim + Thiram (1.1) @ 0.1%.

Ascochyta blight

2. Grey mould

Causal organism: *Botrytis cinerea*

Symptoms

- The symptoms appear as gray or dark brown lesions on the leaflets, branches and pods.
- Small water soaked spots are produced on leaflets. Spots on infected leaves become dark brown.
- Lesions become covered with hairy sporophores of the fungus under high humidity and low temperature conditions.
- Growing tips and flowers are particularly susceptible to infection. The affected plant turns yellow and shedding of infected flowers take place.

Management

- Treat the seeds with Carbendazim (50WP) at the rate of 0.1% or Thiram (75 WP) at the rate of 0.3% or Carbendazim + Thiram (1:1) at the rate of 0.1%.
- Spray the crop with Carbendazim @ 0.1%.
- Crop rotation with non host crops.
- Use resistant varieties e.g PBG-1 and Gaurav

Grey mould of chickpea

3. Dry root rot

Causal organism: *Rhizoctonia bataticola*

Symptoms

- The disease generally appears around flowering and podding time as scattered dried plants.
- Drooping of petioles and leaflets is confined to top of the plant. Sometimes when the rest of the plant is dry, the top most leaves are chlorotic.
- The tap root is dark, shows signs of rotting, and is devoid of most of its lateral and finer roots.
- The tap root on pulling out is found to be dry and devoid of lateral and fixer roots. The dead root shows shedding of bark and is brittle top wards tip. Very minute, dark brown sclerotia of fungus can be seen below the dark.

Management

- Grow resistant varieties.
- Follow crop rotation and crop sanitation.
- Treat the seeds with Carbendazim (50WP) at the rate of 0.1% or Thiram @ 0.3% or Carbendazim+ Thiram (1:1) at the rate of 0.1%.
- Follow crop rotation with non host crops.
- Seed treatment with biocides like *Trichoderma viride* at the rate of 5g per kg of seed.
- Seed treatment with captan/ thiram/ benlate/ PCNB at 3 g/kg of seed

Dry root rot of chickpea

4. Collar rot

Casual organism: *Sclerotium rolfsii.*

Symptoms

- Affected seedlings turn yellow and collapse, but older seedlings may dry slowly.
- Seedlings when pulled out show rottening at the collar region and down wards. White mycelium coating can be seen on the tap root of dried seedlings.

The disease cycle

All the pathogens survive adverse conditions or between crop seasons as the black sclerotia in soil or from infested seed lots. At appropriate conditions, the sclerotia may germinate by one of two means. Sclerotia can germinate directly by means of mycelium in the soil and infect chickpea stems. After appropriate conditioning like periods of low temperature and moisture, sclerotia could also germinate by means of apothecia bearing ascospores in asci. Ascospores are blown and spread by wind and land on stems or flowers to initiate new infection. Within season spread of the disease may occur directly from infected plant to healthy plants througn direct contact under dense crop canopy conditions.

Management

- Uproot the affected plants at initial stage.
- Use wilt resistant varieties, drain out the excess water from the field.
- Seed treatment with Carboxin or Carbendazim at the rate of 0.1% or seed treatment with Biocides like *Trichoderma viride* at the rate of 5 g/ kg of seed.

Collar rot of chickpea

5. Wilt

Causal organism: *Fusarium oxysporum* f. sp. *ciceri*

Description: *Fusarium oxysporum* is a common soil inhabitant and produces three types of asexual spores; the macroconidia are straight to slightly curved, slender, thin walled usually with three or four septa, a foot-shaped basal cell and a tapered and curved apical cell. They are generally produced from phialides on conidiophores by basipetal division. They are important in secondary infection. The microconidia are ellipsoidal and either have no septum or a single one. They are formed from philades in false heads by basipetal division. They are important in secondary infection. The chlamydospores are globose and have thick walls. They are formed from hypae or alternatively by the modification of hyphal cells. They are important as endurance organs in soils where they act as inocula in primary infection. The telemorph or sexual reproductive stage, of F. oxysporum is unknown.

Symptoms

- The disease attacks both seedlings and grown up plants which dry up retaining normal green colour for some time.
- Uprooted plant show almost intact root system, but when split open show inner black discoloration in roots.

Management

- Uproot the affected plants at initial stage, Use wilt resistant varieties, drain out the excess water from the field. Seed treatment with Carboxin or Carbendazim at the rate of 0.1% or seed treatment with Biocides like *Trichoderma viride* at the rate of 5 g/kg of seed.

wilt of chickpea

Lentil

1. Ascochyta blight

Casual organism: *Ascochyta fabae* f. sp. *lentis*

Symptoms

- Dark brown spots are produced on stem, branches, leaves and pods.
- The lesions on the stems and branches are elongated.
- The characteristic symptoms of the disease are black pinhead-like-pycnidia bodies of the fungus on the spots in the form of concentric rings.
- On tender branches spots rapidly girdle the branch and parts of the plant above the girdle, wilt and dry.
- Grey to tan spots or lesions on leaflets, stems, flowers and pods, with dark margins and often with tiny black fruiting bodies (pycnidia) in the centers.
- Lesions first appear on lower leaflets close to soil surface and spread up the plant canopy. Lesions on stems can girdle the plant resulting in wilting. Leaves may turn brown and die off.

Management

- Use disease free seed.
- Treat the seed with seed Captan @2g/kg seed.
- Destroy the diseased plant debris after harvest to reduce the inoculum of the fungus.

Blight of lentil

2. Anthracnose of lentil

Casual organism: *Colletotrichum lentis*

Description: Anthracnose, which is easily confused with Ascochyta blight, is a serious disease of lentils that can be a problem when an extended period of wet weather occurs after canopy closure.

Symptoms

- Anthracnose causes sunken stem lesions; lesions are generally tan with a darker brown border, with multiple black fungal structures (microsclerotia) in the interior. When lesions girdle the stem, plant mortality results.
- Unlike Ascochyta blight, stem lesions caused by anthracnose first develop at the base of plants.
- Leaf and pod lesions are often tan with a darker brown border, and diseased seeds are often discolored.

Management

1. Crop rotation is key to preventing these disease.
2. Use disease resistant varieties
3. Foliar fungicides efficiently control anthracnose when applied preventatively such as Chlorothalonil and several strobilurin products, which also have a slightly systemic effect. The optimum time 0% application is between the 8 and 12 node stage or approximately one week prior to flowing

Anthracnose

3. Wilt

Casual organism; *Fusarium oxysporum* f. sp. *lentis*

Description

Fusarium wilt is characterized by a reddish-brown discoloration of the vascular system; discoloration is often pronounced in the taproot and may not extend into the stem. The disease plugs the water-transporting tissues of plants, causing wilting and death of plants.

Symptoms

- The fungus is soil borne as well as seed borne.
- The tap root gets infected leading to stunted growth and drying up of leaves.
- Causes more towards flowering and pod formation stage.

Management

- Seed treatment with captan/Thiram or carbendazim @ 0.2%
- Use resistant varieties.
- Seed treatment with fungal biocides viz. *Trichoderma viride*, *Trichoderma harzianum* @ 4g/kg
- Seed treatment with bacterial bioagent viz. *Pseudomonas fluorescens*.

Wilt of lentil

4. Sclerotinia stem rot

Occurrence

This disease is commonly observed when cool, wet weather occurs after canopy closure, but it only reaches economic importance when Sclerotinia disease pressure is high (for instance, when sunflowers or irrigated broadleaf crops are part of the rotational history) and extended periods of wet weather occur. Diseased tissues are initially bleached and later become brown and weathered; when the disease is active, white fungal growth and black fungal resting structures of the causal pathogen can be readily found within the crop canopy on diseased plants.

Symptoms

- Sclerotinia stem rot is characterized by bleached (white) tissues and the deposition of black sclerotia in and on stem tissues.
- When moisture levels are high, cottony white growth can be found on diseased leaves and stems.
- The causal pathogen, *Sclerotinia sclerotiorum*, does not sporulate on disease tissue; if gray spores are observed, Botrytis gray mold may be present.

Management

- Crop rotation reduces the disease.
- Use resistant varieties.
- **Fungicides:** Controlling Sclerotinia stem rot with fungicides can be difficult; the disease is favored by dense canopies, and obtaining satisfactory fungicide coverage when the canopy is dense can be a challenge.

Sclerotinia Stem rot

Field pea

1. Powdery mildew

Casual organism: *Erysiphe pisi*

Nature and Occurrence of the disease: This is a soil-borne disease. The perithecia with their contained asci persist in the soil, where they reach from plant debris, until the following season, when they disintegrate on the host surface by producing germ-tubes, and infect the crop. The spread of disease, generally takes place by means of conidia.

Symptoms

- At the flowering time a white powdery substance appears on the leaves and stems of the plants. These floury patches appear on both sides of the leaves. These may cover large areas of the plant. White floury patches appear on both sides of leaves as well as on the tendrils, pods and stem etc.
- During dry season there is more effective than during wet. As the plant becomes old the colour becomes more of less greyish brown and the leaves stems have a dirty appearance. The powdery patches consist of the mycelium and conidial stage of the fungus.

Management

- Spray the crop with wettable sulphur @0.25%, benomyl/carbendazim @0.1% or Dinocap @0.05%, Seed treatment with carbendazim and Thiram (1:2) @ followed by two sprays of mancozeb @ 0.25% (75 days after sowing).
- The most effective means of controlling powdery mildew is to grow a resistant variety. Leaving 4 years between field pea crops. Powdery mildew does not affect other crops in the rotation.

2. Rust of peas

Casual organisms: *Uromyces fabae*

Nature and recurrence of disease: This is an air-borne disease. Several other hosts are also being infected by this rust. Such hosts are-broad beans, lentil, sweet peas, lathyrus etc. Since the rust infects several hosts, it is being carried from one host to another and may pass from one to another season.

Symptoms

- The conspicuous stage of the rust is the uredial stage. The sori appear on both sides of the leaves and often in circles. The sori are light brown to golden to golden in colour.
- When infection is heavy, the leaves become deformed and withered. On the maturity of the crop the telia appear on the stem and leaves. The telia are dark brown and thinly disturbed. Aecia and pycnia also appear on the same host.
- Aecia are in the form of yellow spots in initial stages which turn brown afterwards pycnia infrequent and inconspicuous.

Management

- Fine sulphur powder has been found effective in the reduction of rust infection while it has not been shown to be practiced in the field.
- The resistant varieties should be grown.

Rust of the pea

3. Downy mildew of peas

Casual organisms: *Peronospora pisi*

Nature and Recurrence of the disease: This is a soil borne disease. The oospores survive in the soil along with plant debris. The oospores are the primary source of inoculum. The spread of disease takes place by means of conidia.

Symptoms

- The fungus forms a downy growth on the under surface of the leaves, in patches of varying size, sometimes covering most of the leaves surface, and of a greyish-violet colour.

- If one observes on the upper side of the leaf of these patches, yellow to brown spots are visible.
- On the under side of the leaf growth of fungus is composed of conidiophores, arising from the mycelium found within the leaf tissues.

Management

- The use of metalaxyl, in seed treatments for downy mildew control, is effective
- Fungicide resistance management strategies should be applied to achieve effective and sustainable downy mildew control in pea crops.
- These strategies should include: alternating phenylamides with chemicals with different modes of action, or using phenylamides in mixtures with non-phenylamide chemicals.
- Appropriate alternative or mixture chemicals include cymoxanil, fosetyl-Al, mancozeb and phosphorous acid. The strategies should be applied for seed treatment fungicide applications to control downy mildew in young crops and for foliar applications of fungicides to control the disease in more mature crops.

Downy mildew of pea

Rapeseed and Mustard

1. Alternaria blight

Causal organism: *Alternaria brassicae* (Berk) Sacc.

Survival, spread and favourable conditions for disease development

The disease is externally and internally seed borne. The pathogen survives through spores or mycelium in diseased plant debris or weed. Moist conditions (more than 70% relative humidity) coupled with warm weather (12-25 °C) and intermittent rains favor the development of disease.

Symptoms

- Initial symptom is the production of small brown circular necrotic spots with yellow halo around them on the leaves.
- The spots increase in size and form light brown to brown or black concentric rings on leaves, stems and pods.
- In severe condition pods turn black in colour and may also rot. Seeds are shriveled and under sized in such pods.

Management

- Grow resistant varieties.
- Treat the seed before sowing with Thiram or Captan @ 2.5 g/kg of seed
- Destroy the disease debris from the previous crop.
- Spray the crop with Blitox or Indofil M-45@250g in 100litres of water per acre at 75 days old crop followed by second spray of Score 25EC @100ml and third spray of Blitox or Indofil M-45 @250g in 100 litres of water per acre at 15days interval.
- Spray the crop with the Mancozeb @ 0.25%.

Alternaria blight of Mustard

2. Rust or white blister

Causal organism: *Albugo candida* (Pers) Kuntz.

Survival, spread and favourable conditions for disease development

- The pathogen survives through oospores in affected host tissues and soil.
- Secondary infection is carried out by sporangia and zoospores which produce new infection.
- Moist (more than 70% relative humidity) coupled with warm weather (12-25 °C) and intermittent rains favors disease development.

Symptoms

- Prominent white or cream yellow scattered pustules appear on the undersurface of the leaves.
- These pustules are raised blisters found on the leaves, stems and floral parts. The blisters burst and liberate a white powder.
- Floral parts are much deformed. The flowers get malformed and become sterile.

Management

- Follow crop rotation with non-cruciferous crops.
- Use Metalaxyl 35 SD for seed treatment at the rate of 3.5 g/kg of seed.
- Spray the crop with Ridomil MZ (75 WP) @ 0.25% at appearance of disease.

Rust or white blisters

- Spray the crop three to four times at 15days interval with 250g of Blitox or Indofil M-45 in 100litres of water per acre at 15days interval.

3. Downy mildew

Causal organism: *Peronospora brassicae* Gaumann

Survival, spread and favourable condition for disease development

- The pathogen survives as oospores on the affected plant tissues in soil and on weed hosts.
- Atmospheric temperature in the range of 10-20 °C and relative humidity>90% RH favors disease development.

Symptoms

- The yellow irregular spots appear on the upper surface of the leaves and white mycelial growth is visible on the under surface of leaf opposite to the spots.

Downy mildew of mustard

- In sever infection the inflorescence is malformed and twisted. No pod formation occurs on the infected inflorescence.

Management

- Use healthy seeds for sowing.
- Spray the crop with Dithane Z-78 with the appearance of disease and repeat the spray 2-3 times at 10 days interval.

4. Powdery mildew

Causal organism: *Erysiphe cruciferanum/Erysiphe polygoni* DC.

Survival, spread and favourable conditions

- The pathogen survives through cleistothecia present in the crop debris in the field.
- High temperature (15-28 °C) coupled with low humidity (<60% humidity) and low or no rainfall with wind favours disease development.

Symptoms

- The typical symptom is the appearance of white floury patches on both sides of the leaves as well as green parts of the plants. Symptoms appear as dirty white, circular, floury patches on either sides of the leaves.
- Under favourable environmental conditions, entire leaves, stems, floral parts and pods are affected.
- The whole leaf may be covered with powdery mass.

Powdery mildew of mustard

Management

- Spray Karathane of the rate of 0.1% at 10 days interval starting from the first appearance of the disease.

Linseed

1. Rust

Causal organism: *Melamposora lini* (Ehrenb.) Lev.

Nature and recurrence of the disease: The disease is air-borne. In India linseed is a winter crop sown in October and harvested in March/April, therefore the teliospores have to over summer and not to overwinter as in United states. The teliospores lose their viability during the summer, and therefore, the teliospores donot play any significant role in the carry over of the disease from year to year under Indian conditions. The rust outbreak takes place, 2-3 months later from the date of sowing in the plains. This suggests that there is no local source of infection in the plains from previous crop.

Symptoms

- Affected plants are very conspicuous owing to the bright orange-coloured uredia with which the leaves, stem capsules are covered.
- At a later stage, as the crop ripens the spots are mostly dark brown turning to black and are crust like. These signs represent telia. The pycnia are inconspicuous. The aecia are orange-yellow which appear on stems and leaves.
- The characteristic symptom is the appearance of yellowish orange pustules on both the surface of leaves.
- The leaves become chlorotic and full off. Later on reddish-brown to black elongated lesions appear on the stems and the young twigs.

Rust of Linseed

Management

- Grow resistant varieties.
- Spray the crop with Dithane Z-78 or Dithane M-45 @ 0.25%.

2. Wilt

Casual organism: *Fusarium oxysporum* Schlecht f.sp. *lini* (Bollex) Syn& Hans.

Nature and Recurrence of disease: This is a seed-borne disease. When the contamination seeds are sown, the seedlings are being infected, resulting in the flax wilt. The secondary infection and spread of disease take place by means of conidia.

Symptoms

- The disease may appear my stage of the crop growth.
- The tops of plants droop; and wilt. The infected plants remain stunted.
- The seedlings are killed before producing the second leaf.
- Young plants may be killed and the older plants become yellowish, remain stunted and finally die.
- The fungus may attack the linseed plant at any stage of its development. Seedlings may be killed even before the formation of second leaves.
- The vascular system is plugged and wilting appears. First leaves become yellow and later they droop down.

Management

- Follow good crop rotation for at least 2 to 3 years.
- Treat the seed with Thiram @ 2g/kg of seed
- Grow resistant varieties.

Wilt of linseed

3. Blight of linseed

Casual organism: *Alternaria lini* Dey.

Nature and recurrence of the disease

The disease is seed-borne, causing primary infection in the field. The secondary infection takes place through wind-borne conidia.

Symptoms

- All the aerial parts of plants are infected but the chief symptoms of the disease appears on the floral parts.
- In the diseased plants the buds fail to open which is later on followed by the appearance of small, dark, black spot near the calyx.
- The spots increase in the size and reach up to the pedicel. The floral parts shrink and rot away and closed buds die out ultimately.

Management

- Seed treatment with Thiram or Captan at the rate 2.5g per kg of seed before sowing.
- Spray the crop with Dithane M-45 or Dithane Z-78 (Zineb) at the rate of 2.5g per litre of water. The first spray should be given with the appearance of initial symptoms sprays may be repeated according to the economic loss.
- Grow the late maturity varieties.

Section - C
Major Weeds of Rainfed Rabi Crops

3

Cereals

List of weeds in *Rabi* season crops

S.No.	Botanical Name	Common/local Name	Family
1.	*Anagallis arvensis*	Pimpernel/Krishnaneel	Primulaceae
2.	*Asphodelus tenuifolius*	Wild onion/Piazi	Xenthorrhoeaceae
3.	*Avena ludoviciana*	Wild oat	Poaceae
4.	*Cannabis sativa*	Indian Hemp/ Bhang	Cannabaceae
5.	*Chenopodium album*	lamb's quarter/bathua	Chenopodiaceae
6.	*Chenopodium murale*	Goose foot/ Khartua	Chenopodiaceae
7.	*Cirsium arvense*	Canada thistle/Kateli	Asteraceae
8.	*Convolvulus arvensis*	Fieldbind weed/Hirankhuri	Convolvulaceae
9.	*Coronopus didymus*	Swinecress/ Pitpapra	Brassicaceae
10.	*Cyperus rotundus*	Yellow nutsedge/ motha	Cyperaceae
11.	*Euphorbia dracunculoides*	Dragon Spurge	Euphorbiaceae
12.	*Euphorbia helioscopia*	Sun spurge	Euphorbiaceae
13.	*Fumaria parviflora*	Fineleaf fumitory/Gajri	Fumariaccae
14.	*Lathyrus aphaca*	Jangli Mattar/Matri	Leguminosae
15.	*Lolium* spp.	Italian ryegrass	Poaceae
16.	*Malva parviflora*	Little mallow	Malvaceae
17.	*Medicago denticulata*	Senji	Fabaceae
18.	*Melilotus indica*	Sweet clover	Fabaceae
19.	*Orobanche* spp.	Broom rape	Orobanchaceae
20.	*Phalaris minor*	Little seed canarygrass/ guli danda	Poaceae
21.	*Poa annua*	Annual blue grass	Poaceae
22.	*Polypogon monspliensis*	Foxtail-grass	Poaceae
23.	*Rumex* spp.	Jangli palak	Polygonaceae
24.	*Sonchus oleraceus*	Bandhania	Asteraceae
25.	*Trigonella* spp.	Blue fenugreek /maini	Fabaceae
26.	*Vicia sativa*	Chatri	Fabaceae
27.	*Vicia hirsuta*	Tiny vetch	Fabaceae

Management of weeds in Cereals

Wheat

Major weeds infesting the wheat fields during *rabi* season are *Phalaris minor* (Sitti), *Avena* spp. (Javi), *Fumaria parviflora* (Gajri), *Anagallis arvensis* (Billi booti), *Convolvulus arvensis* (Field bind weed), *Chenopodium album* (Bathu), *Cynodon dactylon* (Khabbal ghas) etc.

- Broad-leaved weeds associated with wheat crop can be managed with the use of 2, 4-D amine @ 0.75 kg/ha or 2,4-D ethyl ester @ 0.5 kg/ha or carfentrazone-ethyl @ 20 g/ha or metsulfuron-methyl @ 4 g/ha or triasulfuron 20 g/ha in 500-600 L at 28-35 DAS which may coincide with 4-6 leaf stage of the crop.
- Do not spray 2,4-D when wheat has been grown mixed with any legume, dicot crop or mustard.
- Monocot weeds associated with wheat crop can be controlled by spraying isoproturon @ 0.75 kg/ha or clodinafop-propargyl @ 60 g/ha or fenoxaprop-p-ethyl @ 100 ml/ha or pinoxaden @ 40-45 g/ha or sulfosulfuron @ 25 g/ha at 30-35 days after sowing of wheat.
- For the control of both monocot and dicot type of weeds in wheat, apply isoproturon @ 0.75 kg + 2,4-D of ethyl ester @ 0.5 kg/ha or metribuzin @ 200 g/ha during 30-35 DAS.

Management of weeds in Pulses

Chickpea

Weeds dominating the chickpea field during *rabi* season are *Launaea asplenifolia* (Jangli gobhi), *Asphodelus tenuifolius* (Piazi), *Fumaria parviflora* (Billi booti), *Cynodon dactylon* (Khabbal ghas) etc.

- The initial four to eight weeks are most critical for weed competition in chickpea. The first mechanical weeding has been advised at 25-30 DAS and second at 45-50 DAS.
- Pendimethalin @ 1.0 kg/ha or metolachlor @ 1.0-1.5 kg/ha or oxadiazon 0.5-1.0 kg/ha or oxyfluorfen @ 100-150 g/ha or pendimethalin+imazethapyr @ 1.0 kg/ha as pre-emergence are the suitable herbicidal weed management options in chickpea for rainfed conditions.

Field pea

Major weeds competing with the Field pea during the initial stages of crop growth are *Fumaria parviflora* (Gajri), *Anagallis arvensis* (Billi booti), *Convolvulus arvensis* (Field bind weed), *Chenopodium album* (Bathu), *Launaea asplenifolia* (Jangli gobhi), *Asphodelus tenuifolius* (Piazi), *Fumaria parviflora* (Gajri), *Cyperus rotundus* (Dilla/motha) etc.

- In field pea 30 to 45 days after sowing is most critical with respect to crop-weed competition. Pendimethalin @ 1 kg/ha or oxadiazon @ 0.5 kg/ha or alachlor+pendimethalin @ 1.00+0.90 kg/ha or alachlor+isoproturon @ 1.00+0.50 kg/ha as pre-emergence followed by one hand weeding at 30-35 DAS gives very good control of weeds.
- If sufficient availability of labour is there and herbicides are not applied one or two hoeing and weeding between 20 and 40 days after sowing are beneficial with respect to yield.

Lentil

In the lentil 30 to 40 days after sowing is most critical for competition with weeds. Major weeds infesting the lentil field are *Chenopodium album* (Bathu), *Lathyrus spp.* (Chatri-matri), *Vicia sativa* (Ankari), *Melilotus alba* (Senji) and *Cirsium arvense* (Kateli) etc.

- One or two hoeing and weeding between 20 and 40 days after sowing are beneficial for obtaining maximum yield.
- Herbicides like pendimethalin (pre-emergence) @ 0.75-1.0 kg/ha can be used for effective weed control.

Management of weeds in Oilseeds

1. Toria / Mustard / Ghobi-Sarson

In toria/mustard/ghobi-sarson, initial 30-45 days after sowing is critical period of crop weed competition. Uncontrolled weeds in these crops may cause 20-70% reduction in yield. The most common weeds of rape and mustard crop are *Chenopodium album* (bathu), *Lathyrus spp.* (Chatri-matri), *Melilotus indica* (Senji), *Cirsium arvense* (Kateli), *Cyperus rotundus* (Dilla/Motha) and *Fumaria parviflora* (Gajri) etc.

- Under rainfed conditions, one hand weeding/hoeing at 25 DAS, is necessary for effective weed control.
- Hoeing should be done soon after thinning. This, besides creating soil mulch and reducing moisture losses, helps in better growth and proper development of crop plants.

For effective control of weeds in mustard apply pendimethalin @ 1 kg/ha or isoproturon @ 1 kg/ha (pre emergence) dissolved in 500-600 L water. Use of isoproturon can cause toxicity to crop which recovers afterwards. Isoproturan can also be used as post emergence 25-30 days after sowing in raya.

Major weeds of *Rabi* season crops

2. Pictures of major weeds of Rabi season crops

Anagallis arvensis

Asphodelus tenuifolius

Avena ludoviciana

Cannabis sativa

Chenopodium album

Chenopodium murale

Cirsium arvense

Convolvulus arvensis

Coronopus didymus

Cyperus rotundus

Euphorbia dracunculoides

Euphorbia helioscopia

Fumaria parviflora

Lathyrus aphaca

Lolium spp.

Malva parviflora

Medicago denticulata

Melilotus indica

Orobanche spp.

Phalaris minor

Poa annua

Polypogon monspliensis

Rumex spp.

Sonchus oleraceus

Trigonella spp.

Vicia sativa

Vicia hirsuta

Non-crop lands weeds and their management

1. *Parthenium hysterophorus*

Parthenium hysterophorus is an aggressive ubiquitous annual herbaceous weed with no economic importance. It is commonly known as carrot grass, bitter weed, star weed white top etc. *Parthenium* can be managed by following methods:

Parthenium hysterophorus at seedling stage

Parthenium hysterophorus at flowering stage

1. Physical pulling/uprooting of *Parthenium* plants by covering hands with gloves or polythene bags in rainy season.
2. Spraying 10-15% common salt (brine solution) followed by controlled burning
3. Herbicides

a. 2, 4-D sodium, amine or ester @ 0.8-1.0 kg/ha

b. Gylphosate 1-1.5%

c. 2, 4-D @ 1 kg/ha + paraquat @ 0.5 kg/ha

d. Metribuzin 0.5%

e. Seeding with *Cassia spp.*, *Mirabilis jalapa* and planting of marigold may be done wherever possible for providing alternate vegetation that suppress *Parthenium.*

Parthenium compost making should be encouraged by utilizing this weed before flowering for providing nutrient rich manure to the crops.

2. Saccharum spontaneum

It is a perennial grass, growing up to three meters in height, with spreading rhizomatous roots. Leaves are harsh and linear, 0.5 to 1 meter long; 6 to 15 mm wide. *Saccharum spontaneum* can be managed by following methods:

Saccharum spontaneum seedling stage

Saccharum spontaneum flowering plant

1. Cut *Saccharum* plants just above the ground level during December-January.

2. Spray 1 per cent glyphosate solution during May-June on the regenerated fresh growth.

3. Plant either hybrid napier or cenchurs slips during raining season (July-August) to provide a grass cover to the treated area to avoid further encroachment by the weed.

4. Spot treatment of the herbicide on the regenerated growth, if any, may be given in October-November followed by gap filling by either of the above grasses during January-February.

3. Lantana camara

Lantana camara is a highly variable ornamental shrub; it is generally deleterious to biodiversity and human activities.

Lantana camara flowering plant

Lantana camara seed

The management of *Lantana camara* can be done by cutting the bushes near the ground level in the month of April followed by herbicidal spray (Glphyosate 1 per cent) on about 30 cm regenerated growth of *Lantana camara* during June and subsequent planting of fodders like *Setaria* or Napier to provide cover to the treated land so as to avoid further reoccurrence of the lantana bushes.

General voluntary vegetation control

1. Glyphosate 1.0-1.5%+ urea (2%)
2. Paraquat 1 kg/ha
3. Dalapon 5-6 kg/ha

Section - D
Plant Nutrient Functions, Their Deficiency and Toxicity Symptoms in Rainfed Rabi Crops

4

Function of Plant Nutrient and Deficiencies Symptoms

Function of Plant Nutrient

Plants require the right combination of nutrients to survive, grow and reproduce. When plants suffer from malnutrition, they show symptoms of being unhealthy. Too little of any one nutrient can cause problems. Plant nutrients are categorized as macronutrients and micronutrients. Macronutrients are required by plants in relatively large quantities. They include nitrogen, phosphorus, potassium, calcium, magnesium and sulfur. Micronutrients are those elements that plants need in small quantities like manganese, boron, molybdenum, zinc, copper, and iron. Both macronutrients and micronutrients are naturally obtained by the roots from the soil. Plant roots require certain conditions to obtain these nutrients from the soil. First, the soil must be sufficiently moist to allow the roots to take up and transport the nutrients. Sometimes correcting improper watering strategies will reduce nutrient deficiency symptoms. Second, the pH of the soil must be within a certain range for nutrients to be release-able from the soil particles. Third, the temperature of the soil must fall within a certain range for nutrient uptake to occur. The optimum range of pH, moisture and temperature, is different for different types of plants. Nutrients may be physically present in the soil, but not available to plants. A knowledge of soil pH, texture, and history can be very useful for predicting what nutrients may become deficient. Nitrogen, phosphorous, and potash are the main nutrients that are commonly deficient in soils. Keep in mind that each plant variety is different and may display different symptoms.

- Many nutrient deficiencies symptoms look similar.
- Micronutrients are used by plants to process other nutrients or work together with other nutrients, hence a deficiency of one may look like another (for instance, molybdenum is required by legumes to complete the nitrogen fixation process).
- If more than one problem is present, i.e. water stress, insect pressure or disease, occurs simultaneously with a nutrient deficiency, or if two nutrients are deficient simultaneously, the typical symptoms may not occur.

- If any nutrient in excess quantity can be toxic to plants. This is most frequently evidenced by salt burn symptoms. These symptoms include marginal browning of leaves, separated from green leaf tissue by a slender yellow halo. The browning pattern, also called necrosis, begins at the tip and proceeds to the base of the leaf along the edge of the leaf.

A plant's sufficiency range is defined as the range of nutrient necessary to meet the plant's nutritional needs and maximize growth. The width of this range will depend upon individual plant species and the particular nutrient. Nutrient levels outside of a plant's sufficiency range will cause overall crop growth and health to decline due to either a deficiency or toxicity. Nutrient deficiency occurs when an essential nutrient is not available in sufficient quantity to meet the requirements of a growing plant. Toxicity occurs when a nutrient is in excess of plant needs and decreases plant growth or quality. Common nutrient deficiencies are nitrogen (N) and phosphorus (P), with some deficiencies of potassium (K), sulfur (S), boron (B), chloride (Cl), copper (Cu), iron (Fe), manganese (Mn), and zinc (Zn). Micronutrient deficiencies are fairly uncommon with deficiencies of B, Fe, and Zn occurring most often Nutrient toxicity is less common than deficiency and most likely occurs as a result of the over-application of fertilizer or manure. The three basic tools for diagnosing nutrient deficiencies and toxicities are 1) soil testing; 2) plant analysis; and 3) visual observations in the field. Both soil testing and plant are quantitative tests that are compared to the sufficiency range for a particular crop. Visual observation, on the other hand, is a qualitative test and is based on symptoms such as stunted growth or a yellowing of leaves occurring as a result of nutrient stress.

Mobile and Immobile Nutrients

Another step in identifying deficiency symptoms is to determine whether the deficiency is the result of a mobile or immobile nutrient based on where the symptom appears in the plant. Mobile nutrients are nutrients that are able to move out of older leaves to younger plant parts when supplies are inadequate. Mobile nutrients include N, P, K, Cl, Mg and molybdenum (Mo). Because these nutrients are mobile, visual deficiencies will first occur in the older or lower leaves and effects can be either localized or generalized. In contrast, immobile nutrients [B, calcium (Ca), Cu, Fe, Mn, Ni, S and Zn] cannot move from one plant part to another and deficiency symptoms will initially occur in the younger or upper leaves and be localized. Zn is a partial exception to this as it is only somewhat immobile in the plant, causing Zn deficiency symptoms to initially appear on middle leaves and then affect both older and younger leaves as the deficiency develops.

MOBILE NUTRIENTS

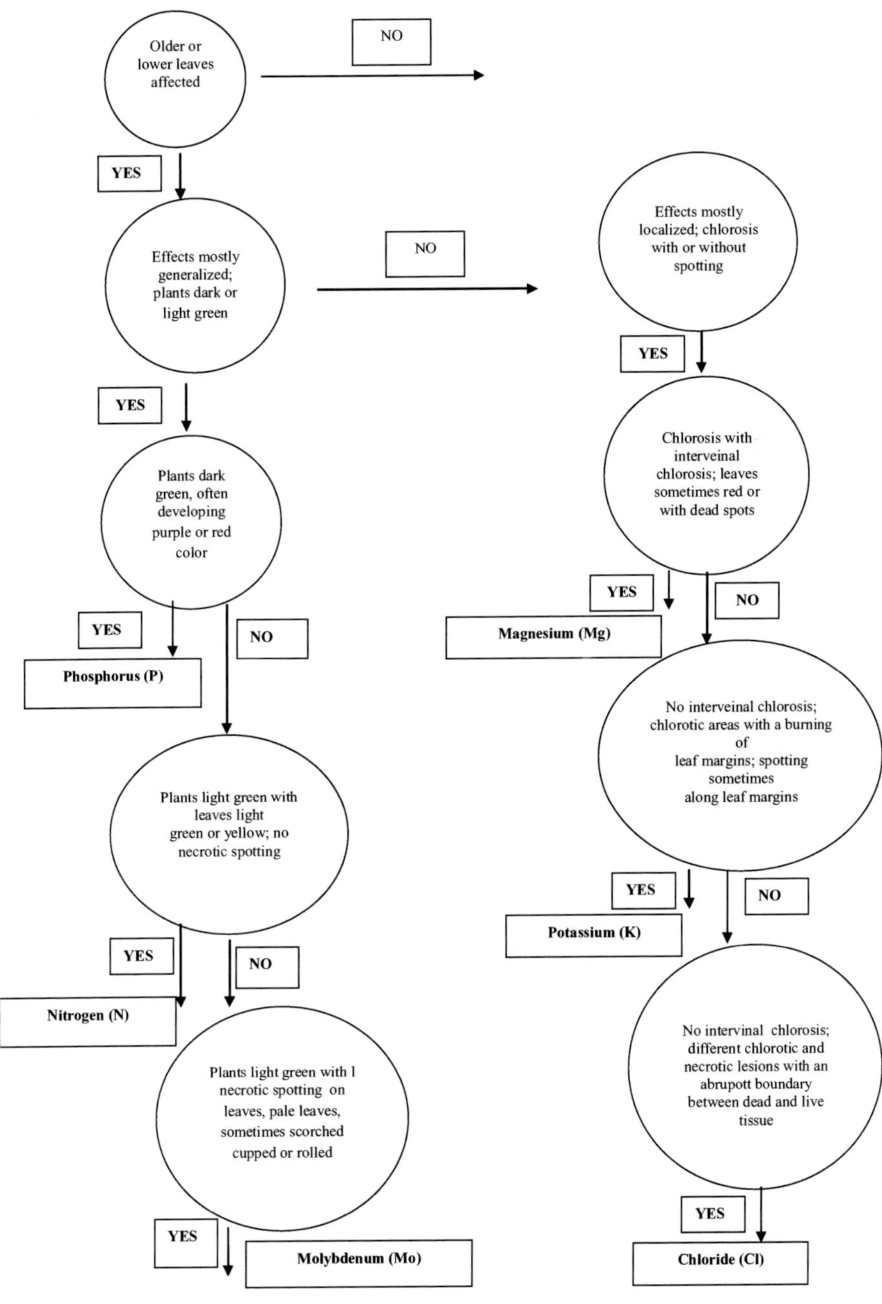

IMMOBILE NUTRIENTS

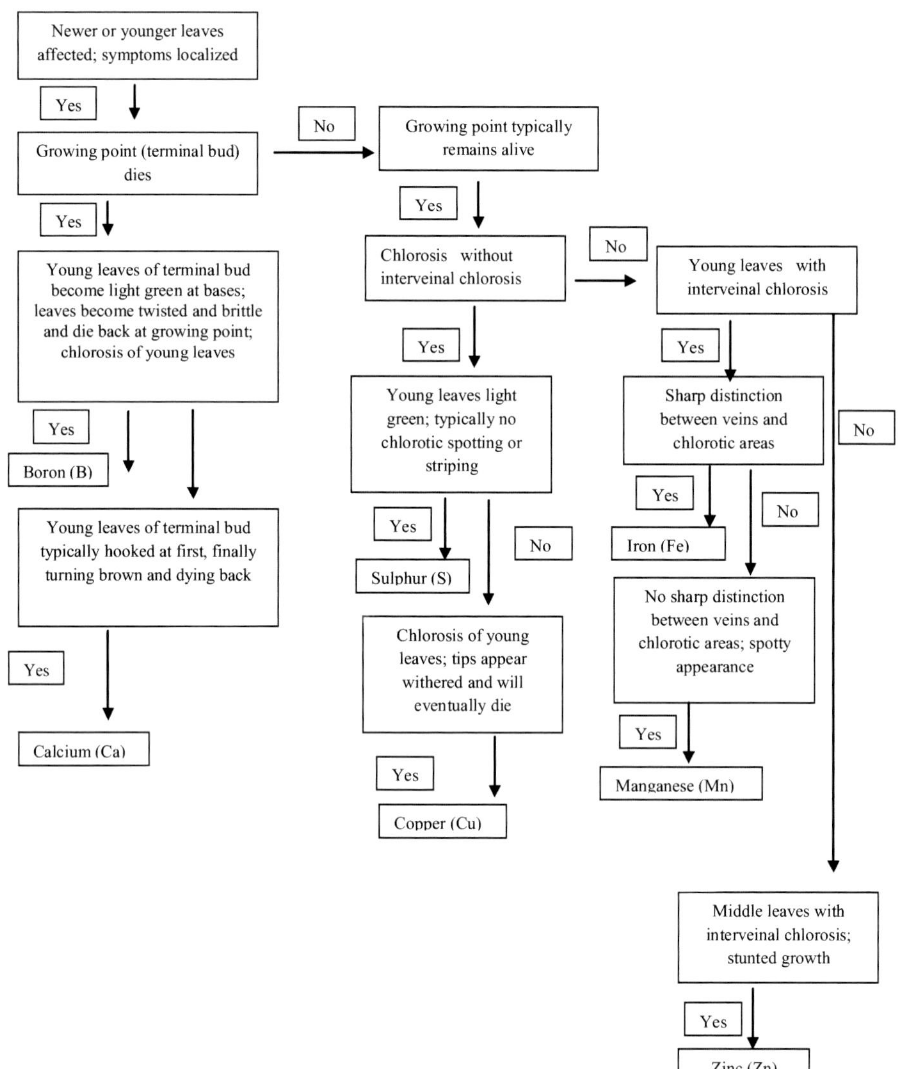

Plant Nutrient Deficiency Terminology

Burning: severe localized yellowing; scorched appearance.

Chlorosis: general yellowing of the plant tissue; lack of chlorophyll.

Generalized: symptoms not limited to one area of a plant, but rather spread over the entire plant.

Immobile: not able to be moved from one part of the plant to another.

Interveinal Chlorosis: yellowing in between leaf veins, yet veins remain green.

Localized: symptoms limited to one leaf or one section of the leaf or plant.

Mobile: able to be moved from one plant part to another.

Mottling: spotted, irregular, inconsistent pattern.

Necrosis: death of plant tissue; tissue browns and dies.

Stunting: decreased growth; shorter height of the affected plants

1. Nitrogen

Symbol: N, available to plants as nitrate (NO_3^-) and ammonium (NH_4^+) ions.

Functions

- Nitrogen is biologically combined with C, H, O, and S to build create amino acids, proteins used in forming protoplasm, the place for cell division and hence for plant growth and development.
- Plant enzymes are made of proteins and nitrogen is needed for all of the enzymatic reactions in a plant.
- It is essential part of the chlorophyll molecule and photosynthesis.
- It is an essential component of several vitamins.
- It improves the quality, succulence of leafy vegetables and fodder crops.

Deficiency symptoms

- The plant becomes light green to light yellow color (chlorosis) appearing first on older leaves, usually starting at the tips. Deficiency of the chlorosis could result in dropping of the older leaves. This is caused by the translocation of N from the older to the younger tissues.
- Stunted crop growth may occur because of reduction in cell division.
- Reduced nitrogen lowers the protein content of seeds and vegetative part.

- Nitrogen deficiency causes the early maturity in some crops, which results in a reduction of yield and quality.

Remedial measures: Deficiency of nitrogen in crop can be ameliorated by applying nitrogenous fertilizers like urea, calcium ammonium nitrate, ammonium chloride as per the recommendation for a crop.

Nitrogen deficiency symptoms in wheat

Nitrogen deficiency symptoms in Gobhisarson

2. Phosphorus

Symbol: P; available to plants as orthophosphate ions (HPO_4^-), $H_2PO_4^-$)

Functions

- In photosynthesis and respiration, phosphorus plays a major role in energy storage and transfer as ADP and ATP (adenosine di- and triphosphate) and DPN and TPN (di- and triphosphopyridine nucleotide). Phosphours is part of the RNA and DNA structures, which are the major components of genetic information.
- Phosphorus is required in large quantities in young cells, such as shoots and root tips, where metabolism is high and cell division is rapid.
- It stimulates root development, flower initiation, and seed and fruit development.
- Phosphorus has a special action on leguminous crops. It induces nodules formation of the crop and rhizobial activity. Thus it helps in fixing more of atmospheric nitrogen in root nodules.
- Phosphorus has been shown to reduce disease incidence in some plants and has been found to improve the quality of certain crops.

Deficiency symptoms

- Phosphorus is needed in large quantities of the early stages of cell division, the deficiency symptom is characterized by slow growth and low yield.
- Phosphours is relatively mobile in plants and can be transferred to sites of new growth, causing symptoms of dark to blue-green colour to appear on older leaves of some plants. Under severe deficiency, purpling of leaves and stems may appear.
- Deficiency of P can cause delayed flowering, maturity, poor seed and fruit development.

Phosphorus Deficiency Symptoms in Gobhisarson

Phosphorus Deficiency Symptoms in Lentil

Phosphorus Deficiency Symptoms in Chickpea

Remedial measures

Phosphorus is sparingly mobile in the soil, its application in the standing crop may not be able to completely meet crop requirement. It is advisable to apply the required dose of phosphorus through proper fertilizer as diammonium phosphate, single superphosphate at the time of sowing in root zone as recommended.

3. Potassium

Symbol: K; available to plants as the ion K^+

Functions

- Unlike nitrogen and phosphorus, potash does not form any vital organic compounds in the plant. However, the presence of potash is vital for plant growth because K is known to be an enzyme activator that promotes metabolism.
- Potash assists in regulating the plant's use of water by controlling the opening and closing of leaf stomata, where water is released to cool the plant. In photosynthesis, potash has the role of maintaining the balance of electrical charges at the site of ATP production.
- Potash promotes the translocation of photosynthates (sugars) for plant growth or storage in fruits and roots.
- It has significant role in ATP production and protein synthesis.
- Potash improves the disease resistance in plants, improve the size of grains and seeds.
- It improves the quality of fruits and vegetables.

Deficiency symptoms

- The most common deficiency symptom is chlorosis along the edges of leaves (leaf scorching). The older leaves show the deficiency symptoms earlier, because K is very mobile in the plant.
- Potash deficient plants show a reduction in photosynthesis.
- In some crops, stems are weak and lodging is common if K is deficient.
- The size of seeds and fruits and the quantity of their production is reduced.

Potassium deficiency symptoms in wheat

Remedial measures

Majority of the soils contain sufficient amount of potassium. Apply potassium fertilizer on soil test basis. In soil testing low in potassium; the recommended dose of the potassic fertilizer (Murate of potash) must be drilled at time of sowing crops.

4. Calcium

Symbol: Ca, available to plants as the ion Ca^{++}

Nutrient function

- Calcium is an important constituent of cell walls and regulates cell wall construction. Calcium deficiency is rare in the presence of calcium carbonates and gypsum in most of agriculture soils.
- Ca promotes root development and growth of plant as it is involved in root elongation and cell division.
- It helps to translocate the sugar in the plant.
- It enhances the nodule formation in leguminous plant and thereby rhizobial activity is increased.

Deficiency symptoms

- Calcium Deficiency symptoms first appear on the younger leaves and leaf tips. The growing tips of roots and leaves turn brown and die.
- Calcium deficiency is not often observed in plants because secondary effects of high acidity resulting from soil calcium deficiency usually limit growth.
- Without adequate Ca, which in the form of calcium pectate, is needed to form rigid cell walls, newly emerging leaves may stick together at the margins, which causes tearing as the leaves expand and unfurl. This may also cause the stem structure to be weakened.
- In some crops, younger leaves may be cupped and crinkled, with the terminal bud deteriorating.

Remedial measures: Apply gypsum (250-500 kg/ha) through soil

Calcium Deficiency Symptoms in wheat

5. Magnesium

Symbol: Mg, available to plants as the Mg^{++} ion

Functions

- Magnesium is an essential constituent of chlorophyll. Several photosynthetic enzymes present in chlorophyll require magnesium as activator.
- It is usually needed by plant in relatively small quantities. Mg helps in formation of chlorophyll and thus imparts dark green colour to leaves. It is indispensable for photosynthesis by plant.
- It plays a part in the production of carbohydrates, proteins and vitamins and helps in the translocation of carbohydrates and fats.

Deficiency symptoms

- Magnesium is a mobile element and part of the chlorophyll molecule. Mg deficiency symptom of interveinal chlorosis first appears in older leaves. Leaf tissue between the veins may be yellowish, brown patches or reddish, the veins of chlorotic leaves remain green.
- Deficiency symptoms may appear on younger leaves and cause premature leaf drop.
- Deficiency symptoms occur most frequently in acid soils and soils receiving high amounts of K fertilizer or Ca in soil.

Remedial measures: Spray of foliar application is more profitable with 1% $MgSO_4$ (Magnesium Sulphate) solution to correct the deficiency.

Magnesium deficiency symptoms in wheat

6. Sulphur

Symbol: S; available to plants as the sulfate ion, SO_4^{2-}

Nutrient Function

- Sulphur is an essential constituent of certain amino acids and proteins. Sulphur is involved in the formation of chlorophyll and thereby encourages vegetative plant growth. It is not a constituent of chlorophyll.
- Sulphur stimulates seed formation and improves the oil content in oilseed crops.
- It promotes the nodule formation on root of leguminous plant.

Deficiency Symptoms

- Sulfur deficiency symptoms initially occur in *young* leaves, causing them to turn light green to yellow (chlorosis). In later growth, the entire plant may be pale green and the characteristic spots or stripes are generally not displayed. Additionally, plants deficient in S tend to be spindly, small and stems are often thin.
- In legumes, the nodulation is poor and nitrogen fixation is reduced.
- Sulphur deficiency symptoms usually appear first on the younger leaves.

Remedial measures: In corps showing deficiency symptoms, gypsum should be applied to soil. Any sulphur containing fertilizer like single superphosphate and ammonium sulphate can be used.

Sulphur deficiency symptoms in wheat

Sulphur deficiency symptoms in pea

7. Manganese

Symbol: Mn; available to plants as Mn^{2+}, Mn^{3+}

Function

- Manganese helps in the synthesis of chlorophyll.
- It helps in protein synthesis in chloroplast and also support the movement of iron in plants.
- Mn acts as catalyst in oxidation and reduction reaction within the tissue.

Deficiency Symptoms

- Chloroplasts (plant organelles where photosynthesis occurs) are the most sensitive of cell organelles to Mn deficiency. As a result, a common symptom of Mn deficiency is interveinal chlorosis in young leaves. However, unlike Fe, there is no sharp distinction between veins and interveinal areas, but rather a more diffused chlorotic spots appear on leaves.
- Mn deficiencies in arable crops are white streak in wheat and interveinal brown spot in barley are also symptoms of Mn deficiency.

Remedial measures: Spray the crop with 0.5% manganese sulphate solution (1kg manganese sulphate in 200 liter of water) 2-3 sprays at weekly intervals on sunny days. Manganese application in soil should be done only after soil testing of that field.

Manganese Deficiency Symptoms in wheat

8. Molybdenum

Symbol: Mo; available to plants as molybdate, MoO_4

Nutrient Function

- Molybdenum enhances the protein synthesis and symbiotic nitrogen fixation in the plant.
- It regulates the activities of several enzymes.

Deficiency symptoms

- Deficiency symptoms resemble those of nitrogen because the function of Mo is to assimilate N in the plant. Older and middle leaves become chlorotic and the leaf margins roll inwards.
- In contrast to N deficiency, necrotic spots appear at the leaf margins because of nitrate accumulation.
- Deficient plants are stunted, and flower formation may be restricted.
- Mo deficiency can be common in nitrogen-fixing legumes.

Remidial measures: Soil application of molybdenum (Ammonium molybdate at recommended dose or foliar spray of sodium molybdate.

9. Copper

Symbol: Cu; available to plants as the ion Cu^{++}

Nutrient functions

- Copper is essential in several plant enzyme systems involved in photosynthesis.
- Cu is part of the chloroplast protein plastocyanin, which forms part of the electron transport chain.

Molybdenum Deficiency Symptoms in wheat

- Cu may have a role in the synthesis of vitamin –A and stability of chlorophyll and other plant pigments.

Deficiency symptoms

- Reduced growth, distortion of the terminal leaves, and possible necrosis of the apical meristem.
- The food synthesis by the process of photosynthesis is hampered.
- In trees, multiple sprouts occur at growing points, resulting in a bushy appearance. Young leaves become bleached, and eventually there is defoliation and die-back of twigs.

Remedial measures: The copper deficiency may be corrected by application of copper sulphate

10. Boron

Symbol: B. Available to plants as borate, H_3BO_3

Nutrient functions

- Boron plays important role in the development and differentiation of tissue, carbohydrates metabolism and translocation of sugar in plants.
- Boron has been shown to promote root growth.

Copper Deficiency Symptoms of Wheat

- Boron is essential for pollen germination and growth of the pollen tube.
- Boron has been associated with lignin synthesis, activities of certain enzymes, seed and cell wall formation.

Deficiency symptoms

- Generally, Boron deficiency causes stunted growth, first showing symptoms on the growing point and younger leaves. The leaves turn yellow or red.
- In many crops, the symptoms are well defined and crop-specific.
- Boron deficiency decreases the rate of water absorption and translocation of sugar in plant.
- Remedial measures for Boron.

11. Zinc

Symbol: Zn; available to plants as Zn^{++}

Functions

- Zinc is an essential component of several enzymes systems which regulates various metabolic reactions in the plants.
- Zn takes part in the synthesis of chlorophyll.
- Zn influences the formation of plant growth hormones, and helpful in reproduction of certain plants.
- Zn has a positive role in photosynthesis and nitrogen metabolism.

Remedial measure: For the correction of deficiency, Borax should be applied on soil test basis and boric acid as foliar spray as recommended.

Boron Deficiency Symptoms in wheat

Deficiency symptoms

- Interveinal chlorosis occurs on younger leaves, similar to Fe deficiency. However, Zn deficiency is more defined, appearing as banding at the basal part of the leaf, whereas Fe deficiency results in interveinal chlorosis along the entire length of the leaf.
- Zn deficiency symptoms in wheat are manifested as the development of light yellow white tissue between the midrib and edge of leaf followed by intervenial chloritic mottling and white to brown necrotic lesion in the middle of the leaf blade.
- In legumes, stunted growth with interveinal chlorosis appears on the older, lower leaves. Dead tissue drops out of the chorotic spots.

Remedial Measures

Soil application is the most efficient way to correct Zinc deficiency. The Zinc deficiency symptoms appear in the crop,zinc sulphate should be top dressed @25kg/ha. Under the acute deficiency condition, 0.5% to 1.0 per cent Zinc Sulphate Solution should be applied.

12. Iron

Symbol: Fe; available to plants as Fe^{2+}, Fe^{3+}

Functions

- Iron is essential for the several enzyme system in plant metabolism (photosynthesis and respiration). The enzymes involved include catalase, peroxidase, cytochrome oxidase, and other cytochromes.

Zinc Deficiency Symptoms in wheat

Zinc Deficiency symptoms in chickpea

- Fe is essential in the synthesis and maintenance of chlorophyll in plants.
- Fe has been strongly associated with protein metabolism.

Deficiency symptoms

- Interveinal Chlorosis of the young or new leaves. Soon after the veins also lose green colour and whole leaf turns yellow. The young leaves may be pale white, due to loss of chlorophyll.
- Usually observed in alkaline or over-limed soils.
- Severe deficiency leaves become dry papery and later turn brown and necrotic.
- Complete leaf fall may occur and shoot can also die.

Remedial measures

Spray the crop with 1.0% ferrous sulphate solution (1kg ferrous sulphate in 100liters of water) at weekly intervals on sunny days till the deficiency is ameliorated

13. Chlorine

Chlorine is involved in osmosis (movement of water or solutes in cells), the ionic balance necessary for plants to take up mineral elements and in photosynthesis. Deficiency symptoms include wilting, stubby roots, chlorosis (yellowing) and bronzing. Odors in some plants may be decreased. Chloride, the ionic form of chlorine used by plants, is usually found in soluble forms and is lost by leaching. Some plants may show signs of toxicity if levels are too high.

Remedial measures: It the soil test sh... chlorine deficiency symptoms, chlorine containing fertilizers should be used, to meet the need. If soil is moderate in potassium, then potassium chloride (Kcl) can be used. Table salt (Sodium chloride, Nacl) should never be used, to increase the soil chlorine level, as excess sodium level can be detrimental to plant health.

Iron deficiency symptoms in wheat

Iron Deficiency Symptoms in Chickpea

14. Nickel (Ni)

Nickel has just recently won the status as an essential trace element for plants according to the Agricultural Research Service Plant, Soil and Nutrition Laboratory in Ithaca, NY. It is required for the enzyme urease to break down urea to liberate the nitrogen into a usable form for plants. Nickel is required for iron absorption. Seeds need nickel in order to germinate. Plants grown without additional nickel will gradually reach a deficient level at about the time they mature and begin reproductive growth. If nickel is deficient plants may fail to produce viable seeds.

Soluble salts like Nickel Sulphate ($NiSO_4$), which contain Ni^{2+}ion are suitable fertilizer to prevent or correct plant Nickel deficiency.

Toxicity

As insufficient nutrient content can cause visual symptoms to occur in plants, so too can an excess. Macronutrient (N, P and K) toxicities most often occur as a result of the over-application of fertilizers or manure. Secondary macronutrient (Ca, Mg and S) toxicities are rare. Micronutrient toxicities can occur and are likely caused by over-application of fertilizer or manure, using irrigation water high in micronutrients or salts, or in areas where soil micronutrient concentrations are abnormally high (i.e. areas exposed to mining activity or high metal minerals in subsoil). In addition, high amounts of non-essential elements such as arsenic (As), cadmium (Cd) and lead (Pb) can be directly toxic to plants and livestock or cause a nutrient imbalance in the plant, in which essential nutrient deficiencies or toxicities may possibly occur.

Macronutrients

Plants with excess N turn a deep green colour and have delayed maturity. Due to nitrogen's positive effect on vegetative growth, excess N results in tall plants with weak stems, possibly causing lodging. New growth will be succulent and plant transpiration high. N toxicity is most evident under dry conditions and may cause a burning effect. Plants fertilized with ammonium (NH^{4+})-based fertilizers may exhibit NH^{4+} toxicity, indicated by reduced plant growth, lesions occurring on stems and roots and leaf margins rolling downward, especially under dry conditions. Excess P indirectly affects plant growth by reducing Fe, Mn and Zn uptake; thus, potentially causing deficiency symptoms of these nutrients to occur (see specific deficiency descriptions). Zn deficiency is most common under excess P conditions. Due to cation imbalance, K toxicity can cause reduced uptake and subsequent deficiencies of Mg, and in some cases, Ca.

Micronutrients

For many crops, the range between deficiency and toxicity is narrower for micronutrients than macronutrients (Brady and Weil, 1999). This is particularly true for B in which the average sufficiency and toxicity ranges for various crops overlap one another: 10-200 ppm (sufficiency range) and 50-200 ppm. B toxicity results in chlorosis followed by necrosis. Symptoms begin at the leaf tip and margins and spread toward the midrib. As the toxicity progresses, older leaves will appear scorched and fall prematurely. Other micronutrients causing potential toxicity symptoms include Zn Cu, Mn, Mo and Ni . Studies suggest excess Cu will displace Fe and other nutrients from important areas in the plant, causing chlorosis and other Fe deficiency symptoms, such as stunted growth, to appear. High Ni concentrations can also cause Fe to be displaced. Interveinal chlorosis may appear in new leaves of Ni toxic plants and growth may be stunted. Mn toxicity symptoms are generally characterized by blackish-brown or red spots on older leaves and an uneven distribution of chlorophyll, causing chlorosis and necrotic lesions on leaves. While Mo toxicity does not pose serious crop problems (crops may appear stunted with yellow-brown leaf discolorations), excess amounts of Mo in forage have been found to be toxic to livestock (Havlin et al., 1999). Zn toxicity is not common, but can occur on very saline soils. Symptoms include 'eaves turning dark green, chlorosis, interveinal chlorosis and a reduction in root growth and leaf expansion. Excess Zn may induce Fe deficiency.

Major symptoms of toxicity of Nutrients

Nitrogen (N

Succulent growth, leaves are dark green, thick and brittle; poor fruit set; excess ammonia can induce calcium deficiency

Phosphorus (P)

Shows up as micronutrient deficiency of Zn, Fe, or Co

Potassium (K)

Causes N deficiency in plant and may affect the uptake of other positive ions such as Mg and Ca

Magnesium (Mg)

Interferes with Ca uptake; small necrotic spots in older leaves; smaller veins in older leaves may turn brown; in advanced stage, young leaves may be spotted

Calcium (Ca)

Interferes with Mg absorption; high Ca usually causes high pH which then precipitates many of the micronutrient so they become unavailable to the plant

Iron (Fe)

Rare except on flooded soils

Boron (B)

Tips and edges of leaves exhibit necrotic spots coalescing into a marginal scorch (similiar to high soluable salts); oldest leaves are affected first; plants are easily damaged by excess application

Zinc (Zn)

Sever stunting, reddening; poor germination; older leaves wilt; entire leaf is affected by chlorosis, edges and main vein often retain more color; can be caused by galvanized metal.

Manganese (Mn)

Reduction in growth, brown spotting on leaves; shows up as Fe deficiency; found under strongly acid conditions

Molybdenum (Mo)

Intense yellow or purple color in leaves; rarely observed

Chlorine (Cl)

Salt injury, leaf burn, may increase succulence

Cobalt (Co)

Little is known about its toxicity symptoms

Nutrient deficiencies and toxicities cause crop health and productivity to decrease and may result in the appearance of unusual visual symptoms. Understanding each essential nutrient's role and mobility in the plant can help determine which nutrient is responsible for a deficiency or toxicity symptom. General deficiency symptoms include stunted growth, chlorosis, interveinal chlorosis, purple or red discoloration and necrosis. Deficiencies of mobile nutrients first appear in older, lower leaves, whereas deficiencies of immobile nutrients will occur in younger, upper leaves. Nutrient toxicity is most often the result of overapplication, with symptoms including abnormal growth (excessive or stunted), chlorosis, leaf discoloration and necrotic spotting. When in excess, many nutrients will inhibit the uptake of other nutrients, thus potentially causing deficiency symptoms to occur as well. As a diagnostic tool, visual observation can be limited by various factors, including hidden hunger and pseudo deficiencies, and soil or plant testing will be required to verify nutrient stress. Nonetheless, the evaluation of visual symptoms in the field is an inexpensive and quick method for detecting potential nutrient deficiencies or toxicities in crops.

15. Sodium (Na)

Sodium is involved in osmotic (water movement) and ionic balance in plants.

16. Cobalt (Co)

Cobalt is required for nitrogen fixation in legumes and in root nodules of non-legumes. The demand for cobalt is much higher for nitrogen fixation than for ammonium nutrition. Deficient levels could result in nitrogen deficiency symptoms.

17. Silicon (Si)

Silicon is found as a component of cell walls. Plants with supplies of soluble silicon produce stronger, tougher cell walls making them a mechanical barrier to piercing and sucking insects. This significantly enhances plant heat and drought tolerance. Foliar sprays of silicon have also shown benefits reducing populations of aphids on field crops. Tests have also found that silicon can be deposited by the plants at the site of infection by fungus to combat the penetration of the cell walls by the attacking fungus. Improved leaf erectness, stem strength and prevention or depression of iron and manganese toxicity have all been noted as effects from silicon. Silicon has not been determined essential for all plants but may be beneficial for many.

References

Atwal, A.S. and Dhaliwal, G.S. 1997. Agricultural Pests of South Asia and their Management. Kalyani Publishers. Ludhiana, India.

B.P. Panday. 1994. AText Book of Plant Pathology. Publisher S. Chand and Company Limited, Ram Nagar, New Delhi

B.R. Bazaya, Dileep Kachroo and Jat, R.K. 2004. Integrated Weed Management in Mustard (*Brassica juncea* L.). *Indian Journal of Weed Science* **36** (3 & 4): 290-292.

Bharat Rajeev, Kachroo Dileep, Sharma Rohit, Gupta M. and Sharma Anil Kumar 2012. Effect of different herbicides on weed growth and yield performance of wheat. *Indian Journal of Weed Science* **44**(2): 106–109.

Brady, N.C., and R.R. Weil. 2002. The Nature and Properties of Soil. Upper Saddle River, N.J. Prentice Hall, Inc. 690 p.

Dhuppar, P.A., Gupta, A. and Rao, D.S. 2013. Chemical weed management in lentil. *Indian Journal of Weed Science* **45**(3): 189–191.

Guide to Symptoms of Plant Nutrient Deficiencies PUBLICATION AZ1106 5/99

http://agritech.tnau.ac.in/crop_protection/crop_prot_crop_insect_pul_bengal%20gram.html

http://farmer.gov.in/imagedefault/ipm/Mustard.pdf

http://ipmworld.umn.edu/ratcliffe-hessian-fly

http://nsdl.niscair.res.in/jspui/bitstream/123456789/493/1/revised%20insect%20pest%20and%20their%20management.pdf

http://www.aicrpchickpea.res.in/plant-path.htm

http://www.nbair.res.in/insectpests/Chromatomyia-horticola.php

http://www.nbair.res.in/insectpests/images/Etiella-zinckenella.php

http://www.northernpulse.com/uploads/resources/618/insect-pests-of-peas-and-lentils-2010.pdf

https://i5k.nal.usda.gov/Mayetiola_destructor

https://www.google.co.in/?gfe_rd=cr&ei=LJBPWJX3KO7x8 AewpomQAw#q= insect+pests+of+wheat

J.A. Silva and R. Uchida. Plant Nutrient Management in Hawaii's Soils, Approaches for Tropical and Subtropical Agriculture.College of Tropical Agriculture and Human Resources, University of Hawaii at Manoa, 2000

Jacobsen, J.S. and C.D Jasper, 1991. Diagnosis of nutrient deficiency in Alfalfa and wheat.EB43,February 1991.Bozeman,Mont.Montana state University Extension.

Jacobsen, J.S. and C.D. Jasper. 1991. Diagnosis of Nutrient Deficiencies in Alfalfa and Wheat. EB 43, February 1991. Bozeman, Mont. Montana State University Extension.

Jones, Jr., J.B. 1998. Plant Nutrition Manual. Boca Raton, Fla. CRC Press. 149 p.

Jones, Jr., J.B. 1998. Plant Nutrition Manual.Boca Raton, fla.CRC pres.149 p.

Malik, R.K. and Singh Samunder. 1995. Littleseed canarygrass (*Phalaris minor*) resistance to isoproturon in India. *Weed Technology* **9**: 419–425.

Masood Ali, Kumar Shiv and Singh, K.P. 2003. Chickpea Research in India. Army Printing Press, Lucknow, India.

McCauley, A 2011. Plant Nutrient Functions and Deficiency and Toxicity Symptoms, Nutrient Management Module No. 9, Montana State University Extension. U.S.

Mengel, K. and E.A. Kirkby. 2001. Principles of plant Nutrition. Netherlands. Kluwer Academic Publishers.849 p.

Mengel, K. and E.A. Kirkby. 2001. Principles of Plant Nutrition. Netherlands. Kluwer Academic Publishers. 849 p.

P. C. Das. Book Manure and fertilizer

Package of Practices for crops of Punjab Rabi 2013-14, 2014-15 Punjab Agriculture University Edited by Dr. H. S. Bajwa pp. 152.

Package of Practices for Rabi Crops. 2007. Sher-e-Kashmir University of Agriculture Sciences and Technology-Jammu, Directorate of Extension Education.

Patel, B.D., Patel, V.J., Patel, J.B. and Patel, R.B. 2006. Effect of fertilizers and weed management practices on weed control in chickpea *(Cicer arietinum* L.) under middle Gujarat conditions. *Indian Journal of Crop Science* **1**(1-2): 180-183.

Poonia, T.C. and Pithia, M.S. 2013. Pre- and post-emergence herbicides for weed management in chickpea. *Indian Journal of Weed Science* **45**(3): 223–225.

Rana, M.C., Kumar, N., Sharma, A. and Rana, S.S. 2004. Management of complex weed flora in peas with herbicide mixtures under Lahaul valley conditions of Himachal Pradesh. *Indian Journal of Weed Science* **36** (1 & 2): 68-72.

Ratnam, M., Rao, A.S. and Reddy, T.Y. 2011. Integrated weed management in chickpea (*Cicer arietinum* L.). *Indian Journal of Weed Science* **43** (1&2): 70-72.

Singh, G., Singh, V.P., Singh, M. and Singh, R.K. 2003. Effect of doses and stages of application of sulfosulfuron on weeds and wheat yields. *Indian Journal of Weed Science* **35**(3&4): 183–185.

Singh, S., Malik, R.K. and Singh, V. 1998. Evaluation of fenoxaprop against *Phalaris minor* in wheat. *Indian Journal of Weed Science* **30**(3&4): 176–178.

Srivastava, K.P. 1996. A Textbook of Applied Entomology (Vol. II). Kalyani Publishers, New Delhi, India.

V.K. Nayyar and I.M. Chhibba Nutritional disorders in field crops Punjab Agricultural University, Ludhiana.

Vasantharaj David, B. 2003. Elements of Economic Entomology. Popular Book Depot, Chennai, India.

Verma, R., Nepalia, V. and Kumawat, S.K. 2004. Influence of Weed Control and Sulphur Nutrition on Weed Dynamics and Productivity of Pea *(Pisum sativum* L.). *Indian Journal of Weed Science* **36** (3 & 4): 285-286.

Yadav, R.B., Vivek, Singh, R.V. and Yadav, K.G. 2013. Weed management in lentil. *Indian Journal of Weed Science* **45**(2): 113–115.

About the Authors

Reena completed her B.Sc. (Agriculture) from Banaras Hindu University, Varanasi, M.Sc. (Agril. Entomology) from University of Agricultural Sciences, Dharwad and Ph.D. (Entomology) degree from C.C.S. Haryana Agricultural University, Hisar. She qualified NET conducted by A.S.R.B., New Delhi and CSIR, New Delhi. She started her professional career as Assistant Professor-cum-Junior Scientist from July, 2004 at SKUAST-Jammu and has thirteen years of experience in Research, Extension work and teaching. She is also the recipient of Junior Research Fellowship (ICAR) during MSc. (Ag) degree program and Department of Science and Technology (DST), Government of India, Young Scientist Project Award (2009- 2012) under fast track scheme for young scientists. She is life member of several professional societies and has handled two externally funded project as PI and two as Co-PI. She has handled about ten University funded research projects as PI and Co-PI. She has delivered several expert lectures to the Department Agriculture personnel, farmers, pesticide dealers and other agriculture workers. She has published more than 25 research papers in journals of national and international repute. She has 10 book chapters, two research manuals and more than ten technical bulletins to her credit.

Dr Sonika Jamwal working as Scientist Senior Scale (Plant Pathology) at Advanced Centre for Rainfed Agriculture, Dhiansar, SKUAST-Jammu has done her B.Sc. (Agriculture) from C.C.S University Meerut, M.Sc. (Plant Pathology) from SHIATS-Allahabad and Ph.D. in Plant Pathology from SHIATS from Allahabad.

She is specialized in biological control agents of soil borne fungus. She is working at SKUAST-Jammu since June 2007 and has ten years of experience.

She is life member of several societies and she has handled two projects as P.I. and six as Co. P.I. She has delivered several expert lectures and has so many Radio and doordarshan talks. She has 20 research papers, 10 book chapters, 30 articles in magazines and newspapers, 15 pamphlets 2booklets, 1 book and one manual on plant diseases to her credit.

Professor Anil Kumar graduated in Agriculture form the SKUAST-J&K, obtained his M.Sc. in Agronomy from HPKV-Palampur and Ph.D. from PAU-Ludhiana. He has served the state and SKUAST-J in different capacities including Agriculture Extension Officer, Assistant Professor Agronomy, programme coordinator of KVK and is presently working as Professor and PI, AICRP-Weed management in the Division of Agronomy, Sher-e- Kashmir University of Agricultural Science and Technology-Jammu. He has more than sixty research papers in national and international journals to his credit, two books and more than thirty chapters in various edited books. He was involved in 8 research projects taken up in the Division from time to time funded by various agencies like, NATP (ICAR), HTMM-I (ICAR), AICRP etc. as PI, life member of nine national and four international scientific societies, vice present of one and councillor of two national societies besides being on the panel of referees for various national and international journals. He is extensively involved in teaching UG, PG courses for the last fifteen years having guided a good number of M.Sc. and Ph.D students and has organized two summer schools sponsored by ICAR.

Dr. Jai Kumar has been presently working as Jr. Scientist, Sr. Scale (Agronomy with Specialization in Field crops, Weed Management in Rainfed Agriculture) at Advance Centre for Rainfed Agriculture, Rakh Dhiansar. His research interests entail Field crops under dryland conditions, Alternate Land Use System, Inter Cropping and soil erosion mitigation. He has worked as Co-PI in many research projects (AICRPDA, NICRA, NABARD) and is involved in research, teaching and extension since 2006. He has published 20 research papers in Journals of national and international repute.

Dr. Pradeep K. Rai obtained his Doctoral degree in Agricultural Chemistry from CSJM Kanpur University, Kanpur, U.P. He joined as Jr. Scientist (Soil Science) at Sher-e- Kashmir University of Agricultural Sciences & Technology –Jammu (SKUAST-J) in 2004 and at present working as a senior scientist (Soil Science). Dr. Rai has published more than 70 research papers, 40 book chapters, more than 100 popular articles, manuals and technical bulletins. He is the member of editorial board of many journals and editor of agricultural Hindi magazines. He attended and organized many National and International Seminars/ Symposia/Conferences. Dr. Rai has received many awards i.e. Young Scientist award, Excellence in Research Award, Leadership Award and Gold Medal Award.

Dr. Brinder Singh is working as a Junior Scientist (Soil Science) at Advance Centre for Rainfed Agriculture (ACRA), Rakh Dhiansar, SKUAST Jammu. He is having 10 years experience in research and extension. He handled research assignments as PI and Co-PIs in AICRPDA scheme. His area of research specialization is Integrated Nutrient Management under Rainfed Agriculture. He has more than 12 research papers in national and international journals to his credit.

Dr. R. Puniya graduated in Agriculture form the RAU, Bikaner, obtained his M.Sc. in Agronomy and Ph.D. from GBPUAT, Pantnagar. He also served as post doctoral fellow in conservation agriculture at IARI, New Delhi. He is presently working as Jr. Scientist in AICRP-Weed Management, Division of Agronomy, Sher-e-Kashmir University of Agricultural Science and Technology-Jammu. He has about more than fifteen research papers in national and international journals to his credit, two book, and four book chapters in various edited books. He was involved in 3 research projects taken up in the Division of Agronomy from time to time funded by various agencies. He is also life member of three national scientific societies. He is extensively involved in teaching UG, PG courses for the last five years having guided two M.Sc. students.